Current Topics in Microbiology and Immunology

Ergebnisse der Mikrobiologie und Immunitätsforschung

Volume 47

Edited by

W. Arber, Genève · W. Braun, New Brunswick · F. Cramer, Göttingen · R. Haas, Freiburg · W. Henle, Philadelphia · P. H. Hofschneider, München · N. K. Jerne, Frankfurt · P. Koldovsky, Prague · H. Koprowski, Philadelphia · O. Maaløe, Copenhagen · R. Rott, Gießen · H.-G. Schweiger, Wilhelmshaven · M. Sela, Rehovoth L. Syruček, Prague · P. K. Vogt, Seattle · E. Wecker, Würzburg

Springer-Verlag Berlin Heidelberg GmbH

ISBN 978-3-642-46162-0 ISBN 978-3-642-46160-6 (eBook)
DOI 10.1007/978-3-642-46160-6

 Originally published by Springer-Verlag Berlin Heidelberg New York in 1969

Library of Congress Catalog Card Number 15-12910. Softcover reprint of the hardcover 1st edition 1969

 Title No. 4695

Table of Contents

School of Veterinary Medicine, University of Pennsylvania
Philadelphia, Pennsylvania 19104

The Poxvirus Antigens

WESLEY C. WILCOX, and GARY H. COHEN * **

Table of Contents

I. Introduction

Only a short time has elapsed since the poxvirus literature was reviewed in a most thorough and comprehensive fashion (JOKLIK, 1966). It is the purpose of this communication to focus in somewhat more detail on a single facet of the overall picture, the poxvirus-coded proteins or antigens. The reader will note that not every investigation bearing on this subject has been cited. The preponderance of our effort has been to review the more recent literature most extensively, with a view towards emphasizing what seems to us are emerging patterns of function. We do not apologize for this approach for although it is intrinsically somewhat arbitrary it has the merit of making the subject more manageable. It also should be apparent to the reader that at least one entire area of investigation, the poxvirus hemagglutinins, has not been mentioned. Relatively little has been done in this field in very recent times and, inasmuch as this subject has been adequately reviewed in 1966 (JOKLIK, 1966), there seems to be little profit in repetition at this time.

* U.S. Public Health Service Postdoctoral Fellow 5-F2-Al-23, 801-02, National Institute of Allergy and Infectious Diseases.

** Present Address: School of Dental Medicine University of Pennsylvania, Philadelphia.

II. The Number of Poxvirus Antigens

The genome of vaccinia virus is a DNA molecule with a molecular weight of about 1×10^8. This amount of DNA is theoretically capable of coding for 400 to 500 proteins. It is somewhat doubtful that it may actually do so inasmuch as only a relatively few virus-specific proteins have been identified to date. Those proteins known, or even strongly suspected to be virus-coded may be divided into three groups at the present time: 1) the virus structural proteins contained in mature virions, 2) the so-called soluble antigens, other than enzymes, which are found in large quantity in poxvirus-infected cells, and 3) the viral-induced enzymes which are protein in nature and as such may be regarded as antigens. These enzymes, which have not yet been unequivocally demonstrated to be virus-coded, will be discussed in a separate section.

Much of the work of recent years in this area has been concentrated upon analyzing the entire contents of the infected cell for unique protein antigens arising after infection. The methods employed, with few exceptions, have been immunological, relying largely upon variations of the Ouchterlony immunodiffusion test or, more recently, immunoelectrophoresis. The intrinsic nature of both of these techniques rules out the participation of anything but soluble antigens inasmuch as virions and hemagglutinins do not diffuse through agar. One might expect with this approach to detect not only the virus structural antigens but also any other unique virus-induced proteins whose functions are other than structural. The effectiveness of these procedures relies heavily upon obtaining good antiserum preparations. Since infected cell extracts are heavily contaminated with host-cell proteins it is important that the antiserum preparations do not react with these proteins. Two procedures have been utilized to minimize this problem: 1) antisera made against extracts of infected cells are thoroughly absorbed with normal cell extract to produce sera which react only with unique, virus-induced antigens, or 2) infected cells of one species are employed to make antisera which are then used to assay the virus-induced antigens produced in a cell system of a different species.

The examination of poxvirus infected cell extracts by such procedures reveals anywhere from seven to 20 immunoprecipitins (APPLEYARD and WESTWOOD, 1964; MARQUARDT et al., 1965, 1965a; WESTWOOD et al., 1965; COHEN and WILCOX, 1966; NICOLE and JOLIBOIS, 1964; RODREQUEZ-BURGAR et al., 1966). This disparity which is of no great consequence, most probably reflects variations in the potency of the antisera used, and the degree of interpretation the investigator is prepared to lavish upon poorly visible and often difficult to reproduce precipitin lines. More important, is that the number of virus-induced proteins detectable by these relatively sensitive techniques is of a different order of magnitude from the 400 to 500 theoretically possible proteins coded for by the viral genome. There is no profit at the present stage of knowledge in attempting to compare closely the results of these various studies or to seek a uniform terminology for the various soluble antigens other than to note that such preparations undoubtedly include the classical LS antigens among the family of immunoprecipitin lines (MARQUARDT et al., 1965 a; ROSENKRANZ et al., 1966). The present status of the NP (SMADEL et al., 1942) and LS (SHEDLOVSKY and SMADEL, 1942) antigens, both of which probably represent the collective behaviour of a number of antigens, will be discussed in more detail shortly. On the basis of results now available it would appear that many, perhaps almost all of the most easily recognizable soluble antigens, are virus structural proteins which have been produced in excess. One group reports that eight of the nine soluble antigens detected are virus structural components (MARQUARDT et al., 1965 a), another that seven out of seven easily recognized soluble antigens are viral components (COHEN and WILCOX, 1968), although another report indicates that only seven out of a total of 17 immunoprecipitins found in soluble extracts could be identified in degraded virus particles (WESTWOOD et al., 1965). It is notable in the latter case however, that many of the antigens represent minor, difficult to detect, components of the soluble extract and may exist in virus in such small amounts as to be virtually impossible to visualize by the methods employed. It is possible that some of the immunoprecipitin lines in the soluble-extracts represent virus-induced enzymes, although at least one report indicates one of the most prominent of these enzymes, thymidine kinase, does not form precipitin lines at the concentrations employed (WILCOX and COHEN, 1967). If indeed there are unique viral-induced proteins that are neither enzymes nor structural proteins their presence or function remains to be established with certainty.

A second approach to the detection of virus-specific protein has been to examine those antigenic moieties which are actually incorporated into the mature virions. The prerequisite for such studies are twofold. First, one must have a highly purified virus preparation free from host cell contaminants, and secondly, if one desires to study the antigenic properties of the virion, a procedure to disrupt the virus particle so as to free the internal proteins without destroying the antigenicity of the structural components is required. Mechanical disintegration of purified virus yields eight precipitinogens, all of which correspond to precipitinogens demonstrable among the soluble antigens (MARQUARDT et al., 1965 a). ZWARTOUW et al. (1965) employing a chemical technique which solubilized about 20% of the viral Nitrogen and 5% of the viral DNA identified eight precipitin lines. WILCOX and COHEN (1967) employing a different procedure identified seven immunoprecipitins in the virus particle corresponding immunologically and, in certain ways, physically to seven readily identified soluble antigens. One might summarize to this point then by saying that all of the antigenic moieties thus far identified in the virion have been similarly identified in the soluble antigen fraction whereas the issue of whether there are soluble antigens which have no virus structural or enzymatic role is still very much open.

Employing still another approach, a most definitive study on the structural components of vaccinia virus was carried out recently by HOLOWCZAK and JOKLIK (1967, 1967a). Radioactively labeled vaccinia virus was dissociated to yield the constituent polypeptide chains which were then subjected to electrophoresis in polyacrylamide gel. By such procedures it was shown that the virion is composed of at least 17 polypeptide components, some major in terms of amount and some minor by the same criterion. Inasmuch as these results do not represent an upper limit on the total number of polypeptide chain species in the virus structural proteins one must conclude that the virus particle is very complex and clearly composed of a fairly large number of different protein species.

Present Status of the NP and LS Antigens

"Early" work done primarily on vaccinia virus, revealed two major groups of antigens (SMADEL et al., 1940; CRAIGIE and WISHART, 1936; SMADEL et al., 1942). These consisted of 1) an antigenic fraction extractable from purified virus with alkali which composed 50% of the mass of the virus and contained all of the viral DNA, termed the NP antigen and 2) an antigen found both in the soluble form and on viral particles known as the LS antigen. It would appear from recent investigations however, that the original concepts concerning these antigens are no longer adequate. WOODROOFE and FENNER (1962) have demonstrated quite conclusively that the singular term NP antigen, is no longer applicable in the original sense, in view of their observation that at least two distinctly different antigens are to be found in this fraction. The first of these, termed the group or family antigen, has obvious taxonomic significance inasmuch as it cross-reacts with similar antigens obtained from all other members of the poxvirus family. It would appear that this is an internal viral protein. The NP fraction also contains surface protein(s) which elicit formation of neutralizing antibodies. The so-called LS antigen also is comprised of at least two distinct antigenic moieties (COHEN and WILCOX, 1966; MARQUARDT et al., 1965 a). Both of these antigens appear to be associated with the coat fraction of the virion inasmuch as antisera prepared against very high concentrations of the LS antigen have been reported to elicit production of neutralizing antibody (COHEN and WILCOX, 1968). It is notable in this regard however, that other investigators find that the LS antigens have no capacity to elicit production of neutralizing antibody (WOODROOFE and FENNER, 1962; COHEN and WILCOX, 1966).

III. Sequental Synthesis of Poxvirus Antigens

Evidence that the poxvirus proteins appear in a sequential fashion has been accumulating for some years. LOH and RIGGS (1961) employing cytochemical and fluorescent antibody procedures first observed the LS antigen(s) in the cytoplasm four hours after infection followed by the NP antigen at 6 hours and viral hemagglutinins at 10 hours. Both the LS and the NP antigens preceeded the first appearance of newly synthesized virus at about eight hours. APPLEYARD and WESTWOOD (1964) employing immunodiffusion analysis noted virus-specific precipitin lines as early as $1^1/_2$ hours after infection. By six hours, four to six precipitin lines were detectable, eight to 10 lines by eight hours and up to 20 lines by the end of the multiplication cycle. There is no reason to suppose that each of these lines did not arise from a separate antigen. More recently WILCOX and COHEN (1967) showed that structural viral proteins may be divided into two groups on the basis of molecular weight as determined by molecular exclusion chromatography. Synthesis of the low molecular weight proteins begins about two hours after infection and virtually ceases after four to five hours, whereas synthesis of the high molecular weight antigens commences at about four hours after infection and continues thereafter. Similarly, SALZMAN and SEBRING (1967) identified early proteins whose synthesis beginning two to three hours after infection, virtually ceased at six hours post-infection to be followed by the synthesis of a different set of proteins. Most recently polyacrylamide gel electrophoresis has been applied to follow the sequential appearance of individual virus structural polypeptide units throughout the course of infection (HOLOWCZAK and JOKLIK, 1967a). This latter work clearly reveals the sequence of appearance and switch-off of virus structural units during the course of infection. It has become increasingly obvious, as this line of research has been pursued, that vaccinia virus functions can be divided into "early" functions, expressed before viral DNA replication commences, and "late" functions which are expressed only after viral genome replication has begun. Presumably the information necessary for the expression of "early" functions resides in the genome of the parental infecting virions whereas the information leading to the expression of the "late" function is transcribed only from replicating viral DNA. This division of the infectious cycle is apparent even at the level of viral messenger RNA in which "early" and "late" vaccinia RNAs differ in stability and base sequence (ODA and JOKLIK, 1967). Evidence to indicate that some of the poxvirus antigens are "early" proteins, in the sense described above, is considerable. APPLEYARD and WESTWOOD (1964) noted some time ago that the synthesis of some viral antigens appeared to precede the onset of viral DNA synthesis by at least one hour. The most convincing evidence for the existence of "early" poxvirus proteins however, has been the demonstration by a number of investigators that production of some viral antigens proceeds under conditions where viral DNA synthesis is not permitted to take place (BECKER and JOKLIK, 1964; ROSENKRANZ et al., 1966; SALZMAN et al., 1964; SHATKIN, 1963; LOH and PAYNE, 1965a; ODA, 1963). Hydroxyurea and 5-fluorodeoxyuridine (FUdR) both specific inhibitors of DNA synthesis were employed in these experiments. In addition, APPLEYARD et al. (1965) have demonstrated by means of immunodiffusion that five viral antigens are synthesized in the presence of isatin-β-thiosemicarbazone, a chemical which affects the ability of late messenger RNA to express itself normally without inhibiting the normal function of early viral messenger (WOODSON and JOKLIK, 1965). These data have led to a more extensive examination of "early" and "late" functions with regard to viral protein synthesis. By combining the technique of immunodiffusion and radioautography SALZMAN and SEBRING (1967) detected five "early" virus-coded proteins arising two to three hours after infection. Under ordinary conditions the synthesis of these "early" antigens, presumably coded for by the input viral genome, stops by the sixth hour. If, however, the synthesis of viral DNA is blocked, these "early" proteins continue to be synthesized. It would appear, therefore, that inhibition of viral DNA replication inhibits a switch-off mechanism which normally operates to block synthesis of "early" proteins beyond 6 hours after infection. At later times in the normal cycle a new spectrum of immunologically distinct viral proteins is made (the "late" proteins). It would appear that the bulk of the viral structural proteins is late in origin, inasmuch as only 15 % of the protein found in mature virus was "early" protein (SALZMAN and SEBRING, 1967). HOLOWCZAK and JOKLIK (1967a) studied the time course of synthesis of structural vaccinia virus proteins by pulse-labeling infected cells at intervals and determining the degree of labeling of individual polypeptide chains of progeny virus after polyacrylamide gel electrophoresis. By this method these investigators identify three "early" polypeptide components, recognizable also by the fact that they are synthesized in the presence of cytosine arabinoside which inhibits viral DNA replication. In view of the fact that these investigators recognize a minimum of 17 polypeptide components in mature virions (HOLOWCZAK and JOKLIK (1967a) it would appear that at least 14 of these components are "late" in origin (coded for by the replicating viral genome). Two of the three early polypeptides are subject to switch-off, although this phenomenon does not seem to be absolute. COHEN and WILCOX (1967, 1968) were able to identify three "early" and four "late" virus structural proteins by immunological procedures. Inhibition of viral DNA synthesis with FUdR or hydroxyurea prevented synthesis of the "late" proteins but had no effect on the synthesis of the "early" antigens.

IV. Location of the Poxvirus Antigens within the Virion

It is possible, on the basis of accumulated data, to approximate the position of some of the virus structural proteins within the virion. COHEN and WILCOX (1968), employing EASTERBROOK'S method (1966) for the chemical degradation of purified virus into core and coat fractions, find that the three "early" virus structural proteins identified in their study appear to be internal proteins. This interpretation is based upon two findings: 1) the "early" antigens were found only in the core fraction of the chemically degraded virus, and 2) the "early" proteins, although immunologically active do not elicit formation of neutralizing antibody. It is notable that MARQUARDT et al. (1965a) employing a different virus fractionation procedure report also that three antigens appear to be associated with the virus nucleoid material. COHEN and WILCOX (1968) further noted that the "late" structural proteins may be subdivided into two fractions on the basis of physico-chemical differences. The first of these fractions, comprising at least two antigens, does not elicit formation of neutralizing antibody. One of these antigens is associated with the core fraction of the virion, the other is found in the coat fraction and presumably occupies an intermediary subsurface position. The second fraction of the "late" proteins, also comprised of at least two protein components, corresponded immunologically to the classical LS antigen(s) and, when in sufficiently high concentration, elicited production of neutralizing antibodies. It would appear, therefore, that these antigens represent coat components of the virion and indeed are found only in the outer (coat) fraction of the partially degraded virion. MARQUARDT et al. (1965 a) have made a similar observation reporting that antigene immunologically identical to the LS antigen(s) were recovered from what appeared to be the purified protein layer of the virion. Still more recently HOLOWCZAK and JOKLIK (1967, 1967a) employing polyacrylamide gel electrophoresis of pulse-labeled, purified virions combined with the controlled chemical fractionation of the virus particles have been able to establish an approximate location within the virion of a number of "early" and "late" virus structural polypeptide components. Of the three "early" polypeptides identifiable by these procedures two are found in the virus core and the third in the virus coat fraction. The "early" core components are subject to switch-off although this phenomenon does not appear to be complete. In addition to the two early components the cores were shown to contain a "late" protein which represented the major core component. With the exception of the single "early" polypeptide the virus coat appears to be comprised of "late" proteins.

The situation with regard to the NP antigen, which is in essence a subviral particle, is less clear. This fraction comprises 50% of the mass of the virion and contains all of its DNA (SMADEL et al., 1942). There are undoubtedly several distinct proteins in this fraction (ZWARTOUW et al., 1965) but the harsh method of extraction appears to destroy the antigenicity of all except a component which reacts in precipitin tests with convalescent antisera to a wide range of poxvirus and a component which is responsible for the production of a low titer homologous neutralizing antibody (WOODROOFE and FENNER, 1962). It is worth noting, however, that the neutralizing capacity of anti-NP sera can be removed by absorption with suspensions of the homologous virus whereas the group reactive component of the sera remains intact (WOODROOFE and FENNER, 1962). It would appear therefore, that the NP preparation contains a surface (coat) antigen and an internal (core) antigen. It has been reported that in the presence of an FUdR concentration which completely inhibits viral DNA synthesis the production of NP antigen, detectable by immunofluorescence, is markedly decreased but not completely prevented (LOH and PAYNE, 1965a). Addition of thymidine mono phosphate (TMP) reversed this effect. This puzzling observation may be explained if one were to assume that the NP complex is comprised of "late" as well as "early" antigens. Failure of the late component(s) to appear after FUdR treatment may then be reflected in an apparent decrease in immunofluorescence.

V. Functions of the Poxvirus Antigens

Virtually nothing is known regarding the function of the various poxvirus proteins whether in the form of soluble antigens or when incorporated into the structure of the virion. There can be little question, however, that these proteins fulfill other than architectural roles. Indeed, in this regard, there is evidence to indicate the existence of a class of non-structural virus-coded proteins in poxvirus infected cells (Holowczak and Joklik, 1962; Westwood et al., 1965). In any event, there is ample reason to suspect the participation of virus-coded proteins in a number of phenomena characteristic of poxvirus infection. Some of the evidence pertaining to this is worthy of mention.

Cytotoxic Effects

A considerable body of literature has accumulated regarding the rapid cytopathogenic effect (cytotoxic effect) elicited by a number of virus types in the absence of viral multiplication (Westwood, 1963). Poxvirus infections are listed among those characterized by this lack of parallelism between virus reproduction and cellular injury. It has been noted, for example, that ultraviolet-inactivated poxviruses are still capable of causing cell death in the absence of virus multiplication (Brown et al., 1959; Appleyard et al., 1962; Hanafusa, 1960). Similarly, mammalian cells treated with sodium azide or Isatin-β-thiosemicarbazone at the time of or prior to infection with rabbitpox exhibited a cytotoxic effect despite the fact that replication of virus did not occur (Appleyard et al., 1962). Appleyard et al. (1962) and Westwood (1963) note that in most instances where cytotoxic effects occur, some viral antigens are made in the infected cells. Thus, it is apparent that although later stages of virus synthesis are prevented some early events proceed. It is probable that the cytotoxic effect involves the partial functioning of the viral genome and, very possibly, the accumulation of certain "early" virus-coded proteins, although the observation that the poxvirus soluble antigens exhibit no cytotoxic effect (Appleyard et al., 1962) does not appear to support the latter part of this hypothesis. It is notable in this regard, however, that protein synthesis must occur prior to the poliovirus-induced cytotoxic event (Bablanian et al., 1965, 1965a).

Switch-Off of Host Cell DNA and Protein Synthesis

Infection of mammalian cells by poxvirus rapidly inhibits synthesis of host cell DNA (Joklik and Becker, 1964; Kato et al., 1964; Kit and Dubbs, 1962; Loh and Payne, 1965) and protein (Shatkin, 1963a). In both cases switch-off appears to be an "early" event inasmuch as it occurs even in the absence of viral DNA synthesis (Salzman and Sebring, 1967; Green and Pina, 1962; Joklik and Becker, 1964). Cessation of host cell DNA synthesis follows infection by ultraviolet-irradiated virus (Joklik and Becker, 1964; Jungwirth and Launer, 1968) and heated (60° C for 15 min) vaccinia virus, although heated cowpox virus exhibited a decreased ability to affect synthesis of cellular DNA (Jungwirth and Launer, 1968). Sambrook and co-workers (1966) also describe a conditional lethal mutant of rabbitpox virus that switches-off host cell DNA synthesis in a non-permissive cell type despite the fact that viral DNA synthesis did not occur in this system. It would appear that the switch-off phenomenon is mediated either by the introduction into the cell of viral protein or by the accumulation of newly-synthesized, "early", virus-coded proteins. The best evidence in favor of the first of these alternatives is the observation that switch-off of host cell DNA by ultraviolet-irradiated virus occurs even in the presence of puromycin (JOKLIK and BECKER, 1964). It would appear, therefore, that switch-off of host cell DNA is not mediated by a protein coded from the viral genome. In contrast LOH and PAYNE (1965) noted some years ago that switch-off requires protein synthesis. Despite the fact that some question remains as to which mechanism is operative the phenomenon appears to be mediated directly or indirectly by viral proteins.

Uncoating

Early in the course of infection the viral genome is freed from its protein coat. This process, termed "uncoating" is required for the full range of expression of the input or parental viral genome. It has been shown, however, that poxvirus cores express some specifically defined areas of their genome prior to the time that they become uncoated. For example, KATES and MCAUSLAN (1967a) have noted that when uncoating is prevented by the addition of an inhibitor of protein synthesis (cyclohexamide), transcription of viral messenger RNA necessary for the synthesis of at least one "early" enzyme (thymidine kinase) proceeds in a normal fashion. Similarly, WILCOX and COHEN (unpublished data) have demonstrated that information necessary for the synthesis of some early structural antigens is transcribed under the same conditions. It is clear, however, that replication of viral DNA is dependent upon completion of the uncoating reaction. It would appear that a poxvirus structural protein, termed the viral inducer protein, induces the formation of a protein, the uncoating protein, which liberates the viral genome from the viral core (JOKLIK, 1966). The recent work of MUNYON et al. (1967) indicating that highly purified vaccinia virus exhibits RNA polymerase activity, reopens the possibility that the uncoating protein is virus-coded. The viral inducer protein is hypothetical in the sense that it has neither been isolated or characterized. Indeed, the only evidence that it is protein in nature is its property of inactivation at relatively low temperature (JOKLIK et al., 1960). Inasmuch as the subject has been extensively and recently reviewed by JOKLIK (1966) there is no reason to discuss this further.

Other Functions

Recently KATES and MCAUSLAN (1967) noted that poxvirus DNA synthesis requires concurrent protein synthesis. The factor, or factors, required, accumulate in the absence of DNA replication and hence may be regarded as "early" protein(s). These materials which appear to be relatively stable do not appear to represent viral-induced thymidine kinase or DNA polymerase by virtue of the instability of its messenger RNA and its stoichiometric rather than catalytic relationship to DNA synthesis. The authors suggest two modes of function: 1) that the protein(s) involved are non-enzymatic structural components of the viral chromosomes necessary for DNA replication but not associated in the chemical aspects of synthesis: or 2) the protein(s) have an enzymatic function but function in a stoichiometric fashion because of an irreversible attachment to the chromosome. Thus, new enzyme(s) could be required for each round of replication.

Production of Neutralizing Antibody

It is well established that immunization of rabbits with soluble antigen preparations elicits formation of virus-neutralizing antibody (APPLEYARD and WESTWOOD, 1964; APPLEYARD, 1961; ROSENKRANZ et al., 1966). The antigens responsible for this activity are virus structural proteins (virus coat protein) which have been produced in excess (MARQUARDT et al., 1965). Immunization of rabbits with soluble antigen preparations confers significant protection against infection, the degree of which, however, is less effective than vaccination with virus (APPLEYARD and WESTWOOD, 1964). The efficacy of vaccination with soluble antigen preparations has, however, not been tested in humans as has been done with adenovirus soluble antigen preparations (KASEL et al., 1964). The poxvirus protein(s) responsible for protection have not been well defined as yet. It has been shown that components of the LS and NP complexes are capable of selectively absorbing neutralizing antibody (CRAIGIE and WISHART, 1934). The reactivity of both anti-LS and anti-NP sera with whole vaccinia virus supports the contention that these are viral surface antigens (SMADEL, 1952). More recently COHEN and WILCOX (1968) presented evidence to indicate that concentrated suspensions of the LS complex are indeed able to stimulate production of neutralizing antibody although earlier investigations show no evidence of this property for LS antigens (ROSENKRANZ et al., 1966; WOODROOFE and FENNER, 1962). It is not unlikely, in view of the structural complexity of the poxvirus virions, that more than one surface component may produce neutralizing antibody.

VI. Separation of Poxvirus Antigens

A number of efforts have been made to isolate and concentrate the various poxvirus antigens with a view towards characterizing individual virus-coded proteins in terms of function as well as structure. Two major lines of approach have been employed. The most direct of these entails either the mechanical or chemical disintegration of preparations of purified virus into component parts. The primary advantage of this procedure is that it avoids the problems of contamination of the viral antigens with host-cell components. Disadvantages lie in the necessity that one must deal with large amounts of purified virus, due to the low concentration in which some antigens occur, and the lability of certain antigens to chemical and mechanical manipulations. A second approach has involved fractionation of the soluble antigen components of the infected cell. The chief advantages of this approach are the relatively large amount of antigens that one may obtain and the possibility of detecting and isolating non-structural virus-coded proteins, other than enzymes, if indeed such proteins exist. The major disadvantages inherent in this approach include the very heavy contamination of the soluble antigen fraction with host cell components and the great antigenic complexity of the soluble antigen fraction.

In 1942 SMADEL et al. (1942) and SHEDLOVSKY and SMADEL (1942) undertook the chemical fractionation of purified vaccinia virus and soluble antigen preparations leading to the discovery of the NP and LS antigens. The procedures involved and the properties of these fractions have been reviewed in detail elsewhere (SMADEL, 1952) and are not described here. More recent work, which has been reviewed in this communication indicates that the LS and NP fractions are considerably more complex than was originally envisaged. In 1964 APPLEYARD et al. obtained some separation of soluble poxvirus antigens by chromatography on DEAE-cellulose, CM-cellulose and calcium phosphate columns. This study was undertaken in an effort to isolate an antigen responsible for the serum blocking effect (depletion of the neutralizing potency of specific antibody following adsorption with soluble antigen preparations). Although unable to isolate a single antigen responsible for the serum blocking effect, chromatography of soluble antigen preparations indicated the presence of at least six precipitating antigens in fractions devoid of serum blocking activity. WESTWOOD et al. (1965) and ZWARTOUW et al. (1965) employed a number of mechanical and chemical techniques to extract soluble immunoprecipitinogens from purified vaccinia particles. Although mechanical disintegration of the virus failed, these investigators were able to extract a number of antigens by trypsin treatment and serial alkaline extraction of purified virions. Tryptic digests which contained roughly 20% of the total viral nitrogen, revealed eight immunoprecipitin lines. Homologous soluble antigen preparations exhibited 17 lines, when tested with the same antisera, of which seven were identical with the trypsin-extracted antigens. In contrast MARQUARDT et al. (1965 a) successfully employed mechanical disintegration of purified virions to release eight precipitinogens all of which corresponded to antigens found in the soluble antigen fraction. These investigators also employed pepsin and subsequent alkaline hydrolysis to release viral antigens. COHEN and WILCOX (1966, 1967) and later MARQUARDT et al. (1967) employed molecular exclusion chromatography, calcium phosphate chromatography, and isoelectric precipitation to effect a physical separation of a number of soluble antigens. By utilizing these procedures singly or in combination it was possible to concentrate and partially purify a number of soluble antigens although a definitive separation of individual antigens has not yet been accomplished. Recently EASTERBROOK (1966) has devised a chemical procedure for the controlled degradation of poxvirus virions that does not appear to cause an extensive destruction of antigens. This procedure, which has been further improved (HOLOWCZAK and JOKLIK, 1967), permits one to fractionate purified virions into core and coat fractions thus making possible an immunological assay of surface and internal proteins (COHEN and WILCOX, 1968). HOLOWCZAK and JOKLIK (1967, 1967a) have employed this procedure in combination with polyacrylamide gel electrophoresis to undertake an extensive study of the structural proteins of virions and cores.

Mechanical fractionation of such gels reveals that the vaccinia virion is comprised of a minimum of 17 polypeptide components. It is unfortunate that the antigenic activity of the protein is destroyed by this procedure which is, otherwise, extremely definitive in that it permits one to determine both the relative mass and approximate location within the virion of each component.

VII. Poxvirus Enzymes

A number of enzyme activities are increased specifically in response to infection with the poxviruses. Among these are thymidine kinase (GREEN, 1962; KIT et al., 1962; McAUSLAN and JOKLIK, 1962), DNA polymerase (GREEN, 1962; GREEN et al., 1964; JUNGWIRTH and JOKLIK, 1965) and a complex of at least three deoxyribonucleases (HANAFUSA, 1961; McAUSLAN, 1965; JUNGWIRTH and JOKLIK, 1965; McAUSLAN and KATES, 1967). The nucleases are differentiated upon the basis of their pH of maximal activity and substrate preference as 1) alkaline DNase (native DNA), 2) neutral DNase (denaturated DNA), and 3) acid DNase (denaturated DNA). These nucleases are induced to varying degrees depending upon the virus strains employed. Although not fully conclusive the evidence that these enzymes are virus-specific, in the sense that the genetic information necessary for their synthesis resides in the viral genome, has become formidable; certainly so in the case of thymidine kinase (KIT and DUBBS, 1965), and DNA polymerase (MAGEE and MILLER, 1967; JUNGWIRTH and JOKLIK, 1965) and to a lesser extent for the nucleases (JOKLIK, 1966; McAUSLAN and KATES, 1967). This evidence has been summarized in detail elsewhere (JOKLIK, 1966; McAUSLAN and KATES, 1967). Inasmuch as these enzymes are immunologically active proteins we must assume, for the time being at least, that they represent poxvirus antigens although their role appears to be other than structural.

The synthesis of thymidine kinase, DNA polymerase, and neutral as well as alkaline DNase are "early" functions both in the sense that they arise early (1—2 hours) in the course of infection (JOKLIK, 1966), and more importantly, that their induction and synthesis does not require synthesis of progeny viral DNA (JOKLIK, 1966). The synthesis of these enzymes is regulated by seemingly similar control mechanisms operating at about four hours after infection. This switch-off mechanism fails to operate when viral DNA synthesis is prevented (McAUSLAN, 1963, 1963a; JOKLIK, 1966). It would appear that this control is encoded in a viral cistron which is expressed only during replication.

Synthesis of acid DNase, in contrast, is a "late" function inasmuch as the enzyme is synthesized only after viral DNA replication begins. The rate of enzyme synthesis is in fact, directly proportional to the amount of viral DNA synthesis (McAUSLAN and KATES, 1967). These data also are consistent with the hypothesis that the genes controlling certain functions are transcribed only from replicating DNA.

In addition to the enzymes listed above it has been noted recently that highly purified vaccinia virions exhibit RNA polymerase activity (MUNYON et al., 1967; KATES and McAUSLAN, 1967b). Several properties of this enzyme differ from those of the RNA polymerase normally found in mammalian cells. For these reasons it was suggested that RNA polymerase is an intrinsic part of the virion and not an enzyme-DNA complex of cellular origin that exists as an impurity in the purified vaccinia preparations. Still more recently MUNYON et al. (1968) have shown that purified vaccinia particles contain a highly active nucleotide phosphohydrolase activity. The virus enzyme(s) converts any of the four common nucleotide triphosphates to their respective diphosphates but does not react with ADP. It was suggested, on the basis of substrate specificity and the response of divalent cations, that the vaccinia enzyme is different from that associated with the microsomes and mitochondria of normal cells. One must suspect, on the basis of these reports, that certain enzymes also function as virus structural proteins. On the other hand, since evidence that RNA polymerase and nucleotide phosphohydrolase are virus-coded enzymes is far from unequivocable, the possibility that these enzymes merely represent host-cell contaminants cannot be entirely discounted.

VIII. Serological Relationships between Poxviruses

On the basis of complement-fixation, precipitin, neutralization, fluorescent-antibody, and hemagglutination-inhibition tests it is widely accepted that the poxviruses are comprised of several subgroups of serologically related viruses as well as a number of agents which are serologically unrelated to each other or to members of the subgroups (DOWNIE and DUMBELL, 1965; FENNER and BURNET, 1957; MAYR, 1959; WOODROOFE and FENNER, 1962). Nevertheless all members of the poxvirus group appear to possess a common or group antigen (WOODROOFE and FENNER, 1962). This antigen is a component of the NP antigen fraction, obtained by alkaline extraction of purified virions and, inasmuch as no serological cross reactivity occurs between viruses of the various subgroups, would appear to be internal protein (WOODROOFE and FENNER, 1962).

The sensitive technique of immunodiffusion has also been employed by a number of investigators to study the antigenic relationships between some of the poxviruses. By these means it is easily possible to differentiate smallpox and vaccinia (NICOLI and JOLIBOIS, 1964; RONDLE and DUMBELL, 1962; MARQUARDT et al., 1965) without resorting to the necessity of employing absorbed antisera as had been done in the past (DOWNIE, 1939). It appears on the basis of such studies that immunodiffusion may be employed to effect a relatively simple and rapid differentiation between members of the various poxvirus subgroups. Immunodiffusion has also been employed to investigate the antigenic composition of the soluble antigen fraction from 10 strains of myxoma (REISNER et al., 1963). It was noted that antigenic differences exist between strains although genetic differences which produce profound physiological effects (virulence versus attenuation) are not necessarily reflected in the antigenic composition of the soluble antigen fraction. It was noted also in this study that fibroma soluble antigen preparations were totally lacking in major antigen components associated with myxoma-infected cells.

There is some evidence for host-induced modification of the soluble antigen fraction. Marquardt et al. (1965) working with extracts from vaccinia-infected mammalian cells report that the relative prominence and, on occasion, even the presence of certain precipitin lines seemed to be dependent upon the cell-type employed. Soluble antigen preparations from vaccinia-infected HeLa cell cultures seemed to be totally lacking in at least one factor found in infected rabbit skin preparations whereas vaccinia-CAM systems were lacking in another.

IX. Summary and Comments

The poxvirus genome contains information for the synthesis of a complex variety of proteins (antigens). The exact number of virus-coded proteins remains in contention. It would appear, however, that this number is of an order of magnitude less than one might expect on the basis of the size of the poxvirus DNA molecule. It has been convenient, primarily on the basis of fractionating infected cells, to divide the virus-specific (presumably virus-coded) proteins into three categories; 1) enzymes which have not yet been unequivocably shown to be virus-coded; 2) soluble-antigens, and 3) virus structural proteins contained in mature virions. This grouping bears little relation to the actual function of these proteins. There is recent evidence, for example, which strongly suggests that certain of the virus-induced enzymes play a virus-structural role (Munyon et al., 1968). In addition, many of the soluble antigens have been shown to be virus structural proteins produced in excess (Cohen and Wilcox, 1968; Holowczak and Joklik, 1967a; Marquardt et al., 1965). On the other hand, a number of the soluble antigens appear to play no virus structural role (Holowczak and Joklik, 1967a; Westwood et al., 1965) although evidence to indicate that these soluble, nonstructural proteins are not virus-induced enzymes has not yet been presented.

It has become very clear that the virus-coded proteins are not all synthesized at the same time during the cycle of infection. Virus-induced enzymes arise between one and four hours after infection (McAuslan and Joklik, 1962; McAuslan, 1963, 1963a; Jungwirth and Joklik, 1965), virus structural proteins are synthesized in a sequential fashion from two hours after infection onward (Marquardt et al., 1965; Joklik and Becker, 1964; McAuslan, 1963a; Holowczak and Joklik, 1967a), and various other virus-coded proteins appear at various times during the course of infection (Applayard et al., 1965; Holowczak and Joklik, 1967a). It is obvious that a mechanism for sequential gene expression operating at the level of transcription or translation or both is involved. Poxvirus functions, as in the case with bacteriophage, can be divided into "early" and "late" functions on the basis of their dependence on viral DNA synthesis. Most of the viral-induced enzymes and some of the virus-structural proteins have been shown to be "early" proteins. On the other hand, at least one enzyme and the bulk of the virus structural proteins are synthesized only after viral genome replication has commenced and are therefore, by convention, "late" proteins. It is possible, by chemical means, to dissociate purified vaccinia virions into core (consisting of the intact genome and internal proteins) and coat fractions (EASTERBROOK, 1966; COHEN and WILCOX, 1968; HOLOWCZAK and JOKLIK, 1967). Employing this technique in combination with others (COHEN and WILCOX, 1968; EASTERBROOK, 1966; HOLOWCZAK and JOKLIK, 1967a) one may approximate the position of various structural proteins within the virion. With the exception of one "late" protein, the core of the virus appears to consist of "early" proteins (HOLOWCZAK and JOKLIK, 1967a; COHEN and WILCOX, 1968). As one might expect, since these are internal proteins, none of these "early" antigens elicit synthesis of neutralizing antibody (COHEN and WILCOX, 1968). In contrast almost all of the coat proteins are "late" proteins (HOLOWCZAK and JOKLIK, 1967, 1967a; COHEN and WILCOX, 1968) although not all of these antigens elicits synthesis of neutralizing antibody (COHEN and WILCOX, 1968) indicating that some represent sub-surface components of the coat. This is consistent with the known morphological complexity of the poxvirus virions. It would appear also, on the basis of recent evidence, that one or more virus-specific enzymes also reside on, or near the surface of the virion (MUNYON et al., 1968).

There is a singular lack of knowledge regarding the functions of the virus-coded proteins. There is, for example, evidence for the existence of a class of virus-coded, non-structural proteins (HOLOWCZAK and JOKLIK, 1967a; WESTWOOD et al., 1965), the functional role of which is entirely unknown. In addition, it is highly improbable that virus-structural proteins have no role other than dictating host-specificity and protection of the viral genome from nucleases. Although direct evidence for the participation of individual virus-coded proteins in specific aspects of infection is lacking, there is reason to suspect that a number of phenomena are mediated by viral proteins. It would appear, for example, that the cytotoxic effect, characteristic of poxvirus infections, is mediated by input viral protein (APPLEYARD et al., 1962; WESTWOOD, 1963). The rapid switch-off of host-cell DNA and protein synthesis is an early event associated either with the accumulation of early, virus-coded proteins or, more likely, input virus structural protein (JOKLIK and BECKER, 1964; JUNGWIRTH and LAUNER, 1968). The so-called "inducer protein" (JOKLIK, 1966) responsible for initiation of the final stages of the uncoating event appears to be a virus-structural component. In addition, there is evidence for the participation of a non-enzymic "early", virus-coded protein in the processes leading to the replication of viral DNA (KATES and MCAUSLAN, 1967).

It is possible by a combination of physical and chemical techniques to partially fractionate the soluble poxvirus antigens. One may, by these techniques, obtain and concentrate individual fractions, each of which contain only a few antigenic moieties against which specific antisera can be produced. It is possible also to degrade purified virus preparations into component protein or polypeptide units. Unfortunately, the complete fractionation of the virion largely destroys the antigenic integrity of the component proteins. One may, by less rigorous methods, partially degrade purified virions without sacrificing antigenicity. These procedures, however, yield complex fractions and would require very large amounts of virus to gain any hope of isolating individual antigens. It would appear, however, that the possibility of obtaining individual virus-coded proteins (antigens) is easily within reach. Once accomplished this may permit one to study the functional as well as the physical characteristics of such proteins.

Most of the authors' personal observations are based on work supported by the National Institute of Allergy and Infectious Diseases Public Health Service research grant AI-06422-04.

References

APPLEYARD, G.: An immunizing antigen from rabbitpox and vaccinia virus. Nature (Lond.) **190**, 465—466 (1961).

— V. B. M. HUME, and J. C. N. WESTWOOD: The effect of thiosemicarbazones on the growth of rabbitpox virus in tissue culture. Ann. N.Y. Acad. Sci. **130**, 92—104 (1965).

—, and J. C. N. WESTWOOD: The growth of rabbitpox virus in tissue culture. J. gen. Microbiol. **37**, 391—401 (1964).

— — A protective antigen from the poxviruses. II. Immunization of animals. Brit. J. exp. Path. **45**, 162—173 (1964a).

— —, and H. T. ZWARTOUW: The toxic effect of rabbitpox virus in tissue culture. Virology **18**, 159—169 (1962).

— H. T. ZWARTOUW, and J. C. N. WESTWOOD: A protective antigen from the poxviruses. I. Reaction with neutralizing antibody. Brit. J. exp. Path. **45**, 150—161 (1964).

BABLANIAN, R., H. J. EGGERS, and I. TAMM: Studies on the mechanism of poliovirus induced cell damage. I. The relation between poliovirus-induced metabolic and morphological alteration in cultured cells. Virology **26**, 100—113 (1965).

— — — Studies on the mechanism of poliovirus-induced cell damage. II. The relation between poliovirus growth and virus-induced morphological changes in cells. Virology **26**, 114—121 (1965a).

BECKER, Y., and W. K. JOKLIK: Messenger RNA in cells infected with vaccinia virus. Proc. nat. Acad. Sci. (Wash.) **51**, 577—585 (1964).

BROWN, A., S. A. MAYYASI, and J. E. OFFICER: The "toxic" activity of vaccinia virus in tissue culture. J. infect. Dis. **104**, 193—202 (1959).

COHEN, G. H., and W. C. WILCOX: Soluble antigens of vaccinia-infected mammalian cells. I. Separation of virus-induced soluble antigens into two classes on the basis of physical characteristics. J. Bact. **92**, 676—686 (1966).

— — Soluble antigens of vaccinia-infected mammalian cells. III. Relation of "early" and "late" proteins to virus structure. J. Virol. **2**, 449—455 (1968).

CRAIGIE, J., and F. O. WISHART: The agglutinins of a strain of vaccinial elementary bodies. Brit. J. exp. Path. **15**, 390—398 (1934).

— — Studies on the soluble precipitable substances of vaccinia. J. exp. Med. **64**, 819—830 (1936).

DOWNIE, A. W.: The immunological relationship of the virus of spontaneous cowpox to vaccinia virus. Brit. J. exp. Path. **20**, 158—176 (1939).

—, and K. R. DUMBELL: Poxviruses. Ann. Rev. Microbiol. **10**, 237—252 (1956).

EASTERBROOK, K. B.: Controlled degradation of vaccinia virions in vitro: an electron microscopy study. J. Ultrastruct. Res. **14**, 484—496 (1966).

FENNER, F., and F. M. BURNET: A short description of the poxvirus group (vaccinia and related viruses). Virology **4**, 305—314 (1957).

GREEN, M.: Studies on the biosynthesis of viral DNA. Cold Spr. Harb. Symp. quant. Biol. **27**, 219—233 (1962).

—, and M. PINA: Stimulation of the DNA-synthesizing enzymes of cultured human cells by vaccinia virus infection. Virology **17**, 603—604 (1962).

GREEN, M., M. PINA, and V. CHAGORGA: Biochemical studies on adenovirus multiplication. V. Enzymes of deoxyribonucleic acid synthesis in cells infected by adenovirus and vaccinia virus. J. biol. Chem. **239**, 1188—1197 (1964).

HANAFUSA, H.: Killing of L-cells by heat- and UV-inactivated vaccinia virus. Biken's J. **3**, 191—199 (1960).

HANAFUSA, T.: Enzymatic synthesis and breakdown of deoxyribonucleic acid by extracts of L cells infected with vaccinia virus. Biken's J. **4**, 97—110 (1961).

HOLOWCZAK, J. A., and W. K. JOKLIK: Studies on the structural proteins of vaccinia virus. I. Structural proteins of virions and cores. Virology **33**, 717—725 (1967).

— — Studies on the structural proteins of vaccinia virus. II. Kinetics of the synthesis of individual groups of structural proteins. Virology **33**, 726—739 (1967a).

JOKLIK, W. K.: The poxviruses. Bact. Rev. **30**, 33—66 (1966).

—, and Y. BECKER: The replication and coating of vaccinia DNA. J. molec. Biol. **10**, 452—474 (1964).

— G. M. WOODROOFE, I. H. HOLMES, and F. FENNER: The reactivation of poxviruses. I. The demonstration of the phenomenon and technique of assay. Virology **11**, 168—184 (1960).

JUNGWIRTH, C., and W. K. JOKLIK: Studies on "early" enzymes in HeLa cells infected with vaccinia virus. Virology **27**, 80—93 (1965).

—, and J. LAUNER: Effect of poxvirus infection on host cell deoxyribonucleic acid synthesis. J. Virol. **2**, 401—408 (1968).

KASEL, J. A., M. HUBER, F. LODA, P. A. BANKS, and V. KNIGHT: Immunization of volunteers with soluble antigen of adenovirus type 1. Proc. Soc. exper. Biol. **117**, 186—190 (1964).

KATES, J. R., and B. R. MCAUSLAN: Relationship between protein synthesis and viral deoxyribonucleic acid synthesis. J. Virol. **1**, 110—114 (1967).

— — Messenger RNA synthesis by a "coated" viral genome. Proc. nat. Acad. Sci. (Wash.) **57**, 314—320 (1967a).

— — Poxvirus DNA-dependent RNA polymerase. Proc. nat. Acad. Sci. (Wash.) **58**, 134—141 (1967b).

KATO, S., M. OGAWA,and H. MIYAMOTO: Nucleocytoplasmic interaction in poxvirus-infected cells. I. Relationship between inclusion formation and DNA metabolism of the cells. Biken's J. **7**, 45—56 (1964).

KIT, S., and D. R. DUBBS: Biochemistry of vaccinia-infected mouse fibroblasts (strain L-M). I. Effects on nucleic acid and protein synthesis. Virology **18**, 274—285 (1962).

— — Properties of deoxythymidine kinase partially purified from non-infected and virus-infected mouse fibroblast cells. Virology **26**, 16—27 (1965).

— —, and L. J. PIEKARSKI: Enhanced thymidine phosphorylating activity of mouse fibroblasts (strain LM) following vaccinia infection. Biochem. biophys. Res. Commun. **8**, 72—75 (1962).

LOH, P. C., and F. E. PAYNE: Effect of p-fluorophenylalanine on the synthesis of vaccinia virus. Virology **25**, 560—576 (1965).

— — Effect of 5-fluoro-2'-deoxyuridine on the synthesis of vaccinia virus. Virology **25**, 575—584 (1965a).

—, and J. L. RIGGS: Demonstration of the sequential development of vaccinial antigens and virus in infected cells; observations and cytochemical and differential fluorescent procedures. J. exp. Med. **114**, 149—160 (1961).

MAGEE, W. E., and O. V. MILLER: Immunological evidence for the appearance of a new DNA polymerase in cells infected with vaccinia virus. Virology **31**, 64—69 (1967).

MARQUARDT, J., E. FALSEN, and E. LYCKE: Isolation of vaccinial immunoprecipitinogens. Arch. ges. Virusforsch. **20**, 109—118 (1967).

MARQUARDT, J., S. E. HOLM, and E. LYCKE: Immunoprecipitating soluble virus antigens. Arch. ges. Virusforsch. **15**, 178—187 (1965).

— — — Immunoprecipitating factors of vaccinia virus. Virology **27**, 170—178 (1965a).

MAYR, A.: Experimentelle Untersuchungen über das Virus der originalen Schweinepocken. Arch. ges. Virusforsch. **9**, 156—192 (1959).

MCAUSLAN, B. R.: Control of induced thymidine kinase activity in the poxvirus-infected cell. Virology **20**, 162—168 (1963).

— The induction and repression of thymidine-kinase in the poxvirus-infected HeLa cell. Virology **21**, 383—389 (1963a).

— Deoxyribonuclease activity of normal and poxvirus-infected HeLa cells. Biochem. biophys. Res. Commun. **19**, 15—20 (1965).

—, and W. K. JOKLIK: Stimulation of the thymidine phosphorylating system in HeLa cells on infection with poxvirus. Biochem. biophys. Res. Commun. **8**, 486—491 (1962).

—, and J. R. KATES: Poxvirus-induced deoxyribonuclease: Regulation of synthesis and control of activity in vivo; purification and properties of the enzyme. Virology **33**, 709—716 (1967).

MUNYON, W., E. PAOLETTI, and J. T. GRACIE, JR.: RNA polymerase activity in purified infectious vaccinia virus. Proc. nat. Acad. Sci. (Wash.) **58**, 2280—2287 (1967).

— — J. O. SPINA, and J. T. GRACIE, JR.: Nucleotide phosphohydrolase in purified vaccinia virus. Virology **2**, 167—172 (1968).

NICOLE, J., et C. JOLIBOIS: Antigenes solubles des "pox virus". I. Etude des antigenes solubles varioliques et vaccinial. Ann. Inst. Pasteur **107**, 374—383 (1964).

ODA, K., and W. K. JOKLIK: Hybridization and sedimentation studies on "early" and "late" vaccinia messenger RNA. J. molec. Biol. **27**, 395—419 (1967).

ODA, M.: Vaccinia-virus HeLa interaction. II. Effect of mitomycin C on the production of infective virus, complement-fixing antigen, and hemagglutinin. Virology **21**, 533—539 (1963).

REISNER, A. H., W. R. SOBEY, and D. CONOLLY: Differences among the soluble antigens of myxoma viruses originating in Brazil and in California. Virology **20**, 539—541 (1963).

RODREGUEZ-BURGAR, A., A. CHORDI, R. DIAZ, and J. TORMO: Immunoelectrophoretic analysis of vaccinia virus. Virology **30**, 569—572 (1966).

RONDLE, C. J. M., and K. R. DUMBELL: Antigens of cowpox virus. J. Hyg. (Lond.) **60**, 41—49 (1962).

ROSENKRANZ, H. S., H. M. ROSE, C. MORGAN, and K. C. HSU: The effect of hydroxyurea on virus development. II. Vaccinia virus. Virology **28**, 510—519 (1966).

SALZMAN, N. P., and E. P. SEBRING: Sequential formation of vaccinia virus proteins and viral DNA replication. J. Virology **1**, 16—23 (1967).

— A. J. SHATKIN, and E. D. SEBRING: The synthesis of a DNA-like RNA in the cytoplasm of HeLa cells infected with vaccinia virus. J. molec. Biol. **8**, 405—416 (1964).

SAMBROOK, J. F., M. E. MCCLAIN, K. B. EASTERBROOK, and B. R. MCAUSLAN: A mutant of rabbitpox virus defective at different stages of its multiplication in three cell types. Virology **26**, 738—745 (1966).

SHATKIN, A. J.: The formation of vaccinia virus protein in the presence of 5-fluorodeoxyuridine. Virology **20**, 292—301 (1963).

— Actinomycin D and vaccinia virus infection in HeLa cells. Nature (Lond.) **199**, 357—359 (1963a).

SHEDLOVSKY, T., and J. E. SMADEL: The LS antigen of vaccinia. II. Isolation of a single substance containing both L and S activity. J. exp. Med. **75**, 165—178 (1942).

SMADEL, J. E.: Smallpox and Vaccinia. Viral and rickettsial infections of man (T. M. RIVERS, ed.), 2nd ed. Philadelphia: J. P. Lippincott Co. 1952.

— G. I. LAVIN, and R. J. DUBOS: Some constituents of elementary bodies of vaccinia virus. J. exp. Med. **21**, 373—389 (1940).

— T. M. RIVERS, and C. L. HOAGLAND: Nucleoprotein antigen of vaccinia virus. I. A new antigen from elementary bodies of vaccinia. Arch. Path. **34**, 275—285 (1942).

WESTWOOD, J. C. N.: Mechanisms of virus infection, (WILSON SMITH, ed.), p. 255—307. New York: Academic Press 1963.

— H. T. ZWARTOUW, G. APPLEYARD, and D. H. J. TITMUSS: Comparison of the soluble antigens and virus particle antigens of vaccinia virus. J. gen. Microbiol. **38**, 47—53 (1965).

WILCOX, W. C., and G. H. COHEN: Soluble antigens of vaccinia-infected mammalian cells. II. Time course of synthesis of soluble antigens and virus structural proteins. J. Virol. **1**, 500—508 (1967).

WOODROOFE, G. M., and F. FENNER: Serological relationships within the poxvirus groups: an antigen common to all members of the group. Virology **16**, 334—341 (1962).

WOODSON, B., and W. K. JOKLIK: The inhibition of vaccinia virus multiplication by isatin-β-thiosemicarbazone. Proc. nat. Acad. Sci. (Wash.) **54**, 946—953 (1965).

ZWARTOUW, H. T., J. C. N. WESTWOOD, and W. J. HARRIS: Antigens from vaccinia virus particles. J. gen. Microbiol. **38**, 39—45 (1965).

Institute of Virology, Czechoslovak Academy of Sciences, Bratislava, Czechoslovakia

Advances in Rickettsial Research

RUDOLF BREZINA

With 2 Figures

Contents

1. Introduction

The majority of definitions of rickettsiae in the past have generally emphasized that these microorganisms represent a link between bacteria and viruses. Today this view — being, after all, so general that it does not mean anything definite — is a mere anachronism. A great number of findings resulting from systematic investigations of fundamental biological, biochemical and physiological problems and of the structure and immunology of rickettsiae contributed to the controversy surrounding this assumption and undoubtedly supported the view that rickettsiae represent a special group of bacteria and any similarity to viruses is restricted solely to the obligatory nature of their intracellular parasitism. Moreover, the new methodical approaches enabled many generally accepted facts to be revised and verified.

I should like to emphasize that this paper was aimed at a discussion of only some recent findings from the field of rickettsiology. Many of them are preliminary, creating the basis for further investigations; some express a more advanced stage of elaboration and often contradict previously accepted findings and hypotheses. It is my intention to leave out completely the questions of relations between rickettsiae and vectors, discussed in detail in reviews by ŘEHÁČEK (1965), HOOGSTRAAL (1967), BURGDORFER and VARMA (1967). Further valuable material summarizing findings from different rickettsiological fields is represented by reviews of WISSEMAN (1968), BREZINA (1968), HIGASHI (1968), WEISS (1968), PARETSKI (1968), FISET and ORMSBEE (1968), and PHILIP (1968). These are the introductory lectures presented in different sections at the Ist International Symposium on Rickettsiae and Rickettsial Diseases, held in Smolenice near Bratislava, September 25—29, 1967.

2. Taxonomic Remarks. — Newly Recognized Rickettsiae

In 1954, two groups of Japanese investigators isolated independently R. sennetsu from patients clinically diagnosed as cases of infectious mononucleosis and glandular fever (MISAO and KOBAYASHI, 1954; FUKUDA et al., 1954). Other similar strains were subsequently isolated from many cases of this disease in Western Japan. The agent has been tentatively placed in the group of rickettsiae but its exact taxonomic position is not yet clear. No serological relationship to other rickettsial agents has been demonstrated. SHISHIDO et al. (1965) studied the pathogenicity of R. sennetsu in cynomolgus monkeys, Macaca irus. The disease induced resembled infectious mononucleosis in man. To date it seems that this agent is not commonly the cause of mononucleosis in the world and that the disease caused by it is limited to Western Japan (Kyushu district).

In the Rocky Mountain spotted fever group of rickettsiae (subgenus Dermacentroxenus), two new species are now included on the basis of serological findings. The rickettsial strain known as the maculatum agent (PARKER et al., 1939; LACKMAN et al., 1949) is now called Rickettsia parkeri and belongs to subgroup B along with Rickettsia conori. Other rickettsiae nonpathogenic for man and previously known as EM (Eastern Montana) agents (BELL et al., 1963) form a new subgroup D along with Western Montana U rickettsiae described by Price (cited in the paper of BELL et al., 1963). It must be stressed that the results of the scientific efforts of the Hamilton staff of workers represent fundamental contributions to the taxonomy of the Rocky Mountain spotted fever group. Their findings are reviewed by LACKMAN et al. (1965).

The typhus group of rickettsiae (R. prowazeki and R. mooseri) has recently been enriched by a third species for which the name Rickettsia canada was proposed (McKIEL et al., 1967). This new rickettsia was isolated from Haemaphysalis leporispalustris ticks removed from indicator rabbits in the vicinity of Richmond, Ontario. In spite of a common complement fixing antigen with R. prowazeki and R. mooseri, this agent is not identical with them, as is demonstrated by the results of the toxin neutralization test in mice. Additional studies concerning the pathogenicity of R. canada for a variety of animals and two species of ticks were performed by BURGDORFER (1968). Infestation of ticks with the typhus group rickettsiae in nature is of interest in the light of recent assumptions that there may exist a second — extrahuman — reservoir of R. prowazeki in ticks and domestic animals, as first suggested by REISS-GUTFREUND (1956). All relevant findings concerning this new, epidemiologically very important hypothesis are summarized and discussed by PHILIP (1968).

Finally, a new group of rickettsia-like agents was recovered from guinea pigs by BOZEMAN et al. (1968) in attempts to isolate infectious agents from human clinical material. Two strains are in many respects related to the previously described two agents revealed in guinea pigs in the course of isolation experiments by TATLOCK and later by JACKSON et al. (cited by BOZEMAN et al., 1967). All four agents possess a common soluble antigen; on the basis of complement fixation reaction with purified suspensions as well as the cross-immunity test, they can be grouped, however, into 3 distinct antigenic types. No serological relationships were demonstrated between the guinea pig agents and the generally known rickettsiae pathogenic for man.

3. Multiplication of Rickettsiae

It is generally accepted that rickettsiae maintain their morphological integrity during growth and multiplication. Binary fission is considered to be the only mode of replication of rickettsiae. This statement is based mainly on the observation of stained rickettsial preparations in the light microscope, but, as we shall see later, electron microscopy also seems to support this view. Numerous data concerning binary fission are summarized by ZDRODOWSKI and GOLINEVICH (1960). SCHAECHTER et al. (1957) observed transverse binary fission of R. rickettsi grown in tissue culture, using phase contrast microscopy. These authors concluded that the reproductive mechanism of rickettsiae was similar to that of bacteria.

Apart from the observations which support the view that binary fission is the only mode of rickettsial multiplication, there are a few older data supporting the possibility of an "invisible" phase of development in rickettsiae. BURNET (1955) supposed the presence of a "lag phase" in the multiplication of Coxiella burneti. ORMSBEE (1952) studied the growth characteristics of Coxiella burneti in embryonated eggs using the complement fixation test. The growth curve exhibited a lag phase and a period of exponential growth. ROBERTS and DOWNS (1959) in their studies of Coxiella burneti multiplication in L cells and chick fibroblasts found a decrease in the LD_{50} titer by 4 $\log_{10}$ units two hours after inoculation of cell cultures. KORDOVÁ and BREZINA (1963) investigated the multiplication dynamics of the same agent in three different kinds of cells. They found a significant decrease in infectivity during the early stages of infection followed by a marked increase in infectivity.

I wish to add some further data concerning this problem, mostly gained in our laboratory. They can be classified into several groups and concern predominantly Coxiella burneti and to a smaller extent Rickettsia prowazeki. This should be emphasized because a generalization of these findings to include other rickettsial species without pertinent experimental data might be quite misleading.

Observations Using Light and Phase Contrast Microscopy

KORDOVÁ (1958) observed in tissue cultures of yolk sac membranes infected with Coxiella burneti minute, weakly stained pin-point particles on the second and third day after infection. Later on she studied the sequence of events during the course of multiplication of this agent in HeLa, KB, HEp—2 and D—6 cells (KORDOVÁ and KVÍČALA, 1962). In the initial stages of infection, a decrease in the number of rickettsiae within the cells was observed. Phase-contrast microscopy revealed no signs of binary fission. The appearence of new particles was preceded by changes in the cytoplasm and development of diverse structures such as dense amorphous matter, large clearly defined formations, homogeneous large bodies, vacuoles and granular particles.

Filterable Particles of Coxiella burneti

It is well known that Coxiella burneti can pass through porcelain and Berkefeld filters which retain bacteria (Davis and Cox, 1938) — hence its other name Rickettsia diaporica. Stoker (1950) found that this agent did not pass through collodion membranes with a pore diameter of 190 nm. In contrast, Kordová demonstrated "filterable" particles of Coxiella burneti using collodion membranes with a porosity ranging from 132 to 65 nm (Kordová, 1959a). Morphologically typical forms appeared only after 3—4 successive passages of filtrates in the chick embryo yolk sac. Titration of the ultrafiltrates in yolk sacs showed the presence of the infectious agent in high titers. During passage "inclusions" and aggregates of dust-like and granular particles were observed in the cell protoplasm prior to the appearence of typical rickettsiae. The same intracellular formations were seen when studying yolk sac tissue cultures inoculated with filtrates. The typical Coxiellae appeared 14—16 days after infection (Kordová, 1959b). Ticks and their organs cultured in vitro seem to be a very sensitive medium for detection of filterable particles of Coxiella burneti (Kordová and Řeháček, 1959a; Kordová and Řeháček, 1959b). In ticks (Ixodes ricinus L., Dermacentor marginatus Sulz., Haemaphysalis inermis Bir.) infected by the intracelomic route, morphologically typical Coxiellae were seen from the eighth to the eighteenth day post infection in the haemolymph and various organs. In cultured tick organs infected with filtrates, multiplication of Coxiella burneti was not as good and slower.

"Live" filtrates induced complement fixing antibodies in a large percentage of inoculated guinea pigs (Kordová, 1959a). Inactivated "antigens", prepared from yolk sacs after the first and second passage of filterable particles, did not react in the CF test with immune sera. Inactivated filtrates did not induce the formation of specific antibodies, probably because the antigens are confined to morphologically "mature" organisms (Kordová, 1960a).

Filterable particles of C. burneti were found to persist in guinea pigs without any signs of infection and they were passaged serially in them (Kordová, 1960b). However, when young animals were inoculated with spleen suspensions from adult "healthy" animals, manifest infection with typical clinical, serological and microscopic findings developed. Thus, filterable particles contain the genetic material. They are capable of persisting in animals and of developing into mature organisms in a "sensitive indicator host".

The question arises as to the biological importance of filterable particles. Are they a by-product of multiplication or do they represent an obligatory stage in the replication cycle? Kordová assumes that filterable particles represent an intracellular stage of multiplication of Coxiella burneti. If this hypothesis is correct, there must be — at least in the early phases — a mode of replication different from binary fission and involving active participation of the host cell.

Electron Microscopic Observations on the Replication of Coxiella burneti

ROSENBERG and KORDOVÁ (1960, 1962) studied the intracellular forms of Coxiella burneti during the course of its propagation in yolk sac as well as in Detroit-6 cell cultures. Upon examination of ultrathin sections of infected yolk sac tissue cultures, the following structures were observed: (a) several μ large matrices without an outer membrane which were present at 48 hours and later, (b) smaller, about 2 μ large formations without an outer membrane, seen 4—6 days after infection, (c) intermediate forms (300—1,200 nm) with a distinct outer membrane, fine-grained viroplasm and a dense centre. These forms often showed segmentation; (d) morphologically typical elementary bodies. In Detroit-6 cell cultures no Coxiella-like particles were found during the first hours after infection. After 48 hours, homogeneous masses of fine granular material were formed in the cytoplasm of infected cells. No evidence was obtained which indicated that multiplication occurred by binary fission. These results are in sharp contrast to those obtained by ANACKER et al. (1964) who found numerous examples of binary fission in ultrathin sections of infected yolk sacs.

Histochemicl and Fluorescent Antibody Studies

KORDOVÁ and KOVÁČOVÁ (1968) studied the development of Coxiella burneti in monolayer cultures of L cells with the use of infectivity assays, acridine orange staining and immunofluorescence. Aside from an eclipse phase within 2—6 hours after infection, acridine orange staining revealed an increased content of deoxyribonucleic acid in the cytoplasm from 4—6 hours post infection, preceding by about 10—12 hours the appearance of specific antigen as demonstrated by fluorescence. In spite of the large number of Coxiellae present immediately after penetration into cells, the specific antigen disappeared 4—6 hours post infection and new organisms were revealed by immunofluorescence 20 hours p.i., mainly in the paranuclear area.

Studies on Replication of Rickettsia prowazeki

It is remarkable that the morphological changes occurring in cells infected with this agent resemble in many respects those found during the multiplication of Coxiella burneti. They were demonstrated in ticks (KORDOVÁ and ŘEHÁČEK, 1964) and in L cell cultures using light microscopy (KORDOVÁ, 1965), electron microscopy (KORDOVÁ et al., 1965), acridine orange staining (KOVÁČOVÁ and KORDOVÁ, 1966) and immunofluorescence (KORDOVÁ and KOVÁČOVÁ, 1967).

Summarizing the above-mentioned data, there seems to be a "mysterious" phase in early stages of replication of at least Coxiella burneti and R. prowazeki, and consequently there emerge a number of stimulating questions which dedeserve further study.

If the cell is involved in the early stages of rickettsial multiplication in terms of some synthesizing functions, it would be very difficult to understand the extracellular propagation of such a rickettsia as R. mooseri. WEYER (1968) believes that he succeeded in obtaining extracellular growth of this agent in the haemolymph of body lice. According to this author, extra- and intracellular multiplication does not represent a fundamental biological difference and should not be considered an important criterion for classification of these organisms. WEYER'S new R. mooseri strains were shown to correspond in all essential festures to R. quintana, which is known to multiply outside cells and in cell-free media (VINSON, 1966). Binary fission as the mode of multiplication of rickettsiae is accepted also by KOKORIN (1968) who studied various rickettsiae grown in tissue cultures by light microscopy, cine-micrography and electron microscopy. The new findings suggest the existence of two developmental stages in rickettsiae: vegetative, dividing and moving forms on the one hand and resting, immobile forms on the other. During intensive multiplication of rickettsiae under favourable conditions, the majority of them lose their mobility and are transformed into immobile, resting forms. After change of the nutrient medium, transfer of infected cultures, or infection of fresh cells, the resting forms change again into vegetative forms. Resting forms can persist in the cell for a very long period of time, thus providing a possible explanation for latency in rickettsial infections.

4. Structure of Rickettsiae

In terms of chemical, morphological and antigenic composition, rickettsiae are complex organisms, very similar to bacteria. They are Gram-negative, except Coxiella burneti which is Gram-labile (GIMENEZ, 1965). There is little fundamentally new to add to past observations on their morphology as revealed by the light microscope. Only two new findings should be mentioned in this respect. BURGDORFER and ORMSBEE (1968) detected by fluorescent antibody staining hitherto undescribed "flagellated" forms in the "ZRS" strain of R. prowazeki grown in the haemolymph of experimentally infected ticks. On the other hand, BURGDORFER (1968) found "comet-shaped" forms of R. canada in cells of infected ticks.

Many observations on ultrastructure of rickettsiae were made recently. Unfortunately, different preparative techniques were used which has resulted in more or less significant differences in the appearance of intracellular organisms. In spite of this, there are many similarities between various species of rickettsiae. Electron microscopic methods are not yet sufficiently widely used to follow all stages of rickettsial multiplication.

After the recent comprehensive review of the ultrastructure of rickettsiae (HIGASHI, 1968), I should like to mention only a few data regarding this field.

Each rickettsial species has a cell wall and plasma membrane and contains ribosomes and deoxyribonucleic acid in a ground substance.

R. prowazeki is enclosed by a plasma membrane and a cell wall, both showing a tri-layered structure (ANDERSON et al., 1965). Dense ribosome-like granules and electron-lucent spherical structures, as well as DNA-like strands are the main internal features of this organism. Figures of binary fission are often seen. Shkolnik et al. (1966) found two membranes in this agent, the external being undulate and consisting of three layers. The most interesting finding of these authors is the different appearance of the cell wall in various R. prowazeki strains. In strain Z they demonstrated vesicular protrusions from the cell wall in various stages of release from the rickettsial body. This strain exhibited also a distinct internal structure, that is a more concentrated localization of the dense granules and absence of vacuole-like formations. Examining

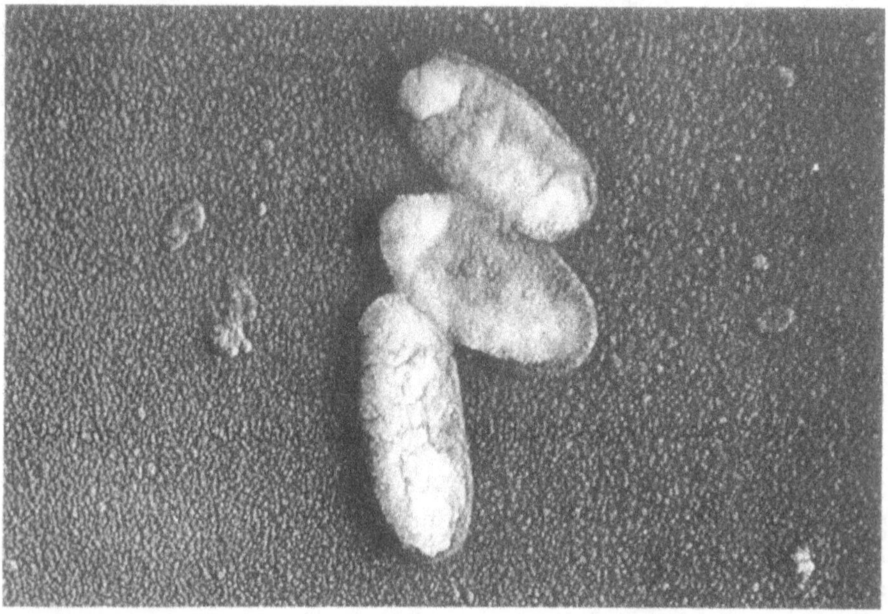

Fig. 1. Coxiella burneti, strain Florián, phase I, shadowed. Enlarged 40,000 ×

the fine structure of R. prowazeki in ultrathin sections of haemocytes of experimentally infected ticks, Bird et al. (1967) demonstrated outer and inner layers of double membranes. The internal structure was shown to be similar to that described by the already cited authors. Anacker et al. (1967) found a five-layered cell wall in R. prowazeki grown in the yolk sac. Apart from these five layers, another one was seen, rather amorphous and faintly stained, which surrounded the sharply defined cell wall. This material possibly represents the soluble antigen. The presence of intracytoplasmic membraneous organelles, observed already by Anderson et al. (1965) were also noted.

The structure of *R. rickettsi*, *R. tsutsugamushi* and *R. sennetsu* was studied by Anderson et al. (1965). R. rickettsi was shown to be similar to R. prowazeki except for the absence of the lucent spherical structures in its ground substance. In the case of R. tsutsugamushi, DNA strands appeared often in a discrete nuclear area. R. sennetsu is surrounded by a tri-layered cell wall and a plasma membrane. The internal structure exhibits prominent ribosomes and irregular patches containing DNA. In the cytoplasm of cells this rickettsia is always enclosed in a vacuole.

The fine structure of R. quintana cultivated in vitro and in the louse was studied by ITO and VINSON (1965). As in other rickettsiae, an outer cell wall and plasma membrane, both trilaminar, were found. Vesicular invaginations of the plasma membrane were noted occasionally. The cytoplasm contained ribosomes and both DNA and RNA could be demonstrated in it.

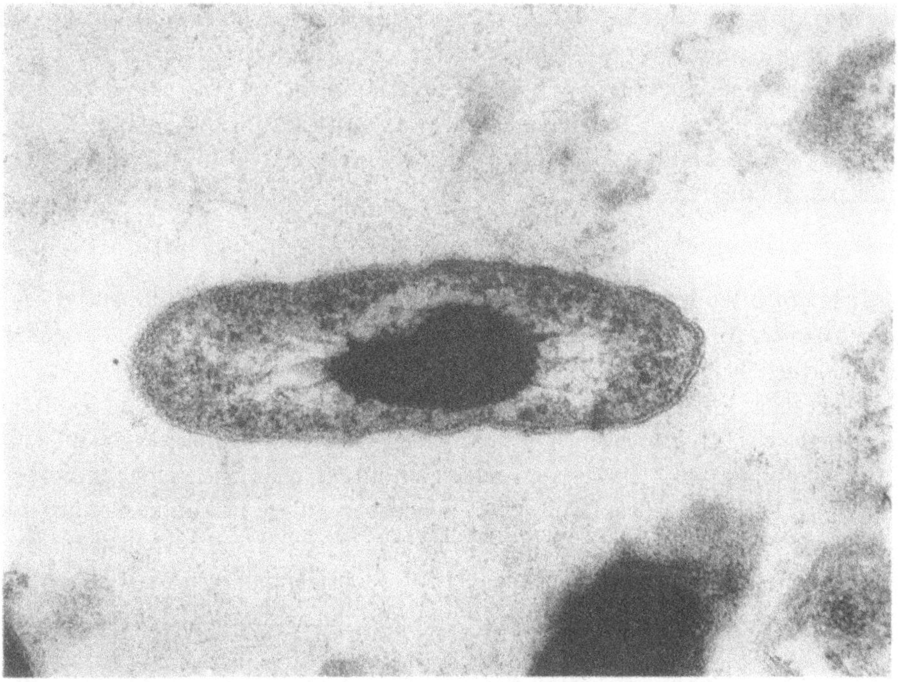

Fig. 2. Coxiella burneti, strain Florián, phase I. Ultrathin section. Note cell wall, granular cytoplasm and typical "central dense body" lying in the centre of a light area (probably nucleoplasm). Enlarged 120,000 ×. (Published with permission of Dr. M. NERMUT, Institute of Virology, Czechoslovak Academy of Sciences, Bratislava)

Finally, several authors tried to elucidate the fine structure of Coxiella burneti. As previously mentioned, ANACKER et al. (1964) followed the development of this agent in the yolk sac. In 3—7 day-old cultures dense fibrillar centers were observed which were surrounded by less dense cytoplasmic material. As in other rickettsiae, several multiple external layers were demonstrated. Moreover, atypical forms were found which did not bind ferritin-labelled antibodies. The nature of these forms remains unknown. Are they artifacts or normal stages of maturation? HANDLEY et al. (1967) studied Coxiella burneti electron microscopically in infected liver and splenic tissue of guinea pigs. They demonstrated large clusters of Coxiellae in the cytoplasm surrounded by a limiting membrane, sometimes multilayered. Within the clusters, located in vacuoles, various forms were detected with a clearly delineated cell wall surrounding a cytoplasmic membrane. Urbach et al. (1968) obtained roughly the same results with regard to the formation of inclusions in guinea pig testes.

Since Coxiella burneti may exist in phase I or II according to the phase variation phenomenon, there arises the question of a possible difference in the structure of the cell wall of these phases. In a preliminary study, Nermut et al. (1967) did not find any difference.

Insufficient data are available concerning the relationship between structure and antigenic composition of rickettsiae and regarding the chemical nature of various structural units. Studies on the chemical and immunological properties of the rigid portion of cell wall in R. mooseri (Schaechter et al., 1957; Wood and Wisseman, 1967a, b), R. prowazeki (Frygin and Siwecka, 1965; Frygin, 1966), and Coxiella burneti (Anacker et al., 1962; Allison and Perkins, 1960) created the basis for further investigations in this field.

5. Metabolism of Rickettsiae

It is not intended to make a complete survey of all metabolic activities of rickettsiae in this chapter. The present knowledge in this field has been reviewed recently by Weiss (1968), Paretski (1968), and Zubok (1966).

Rickettsiae, like other microorganisms, may be studied using intact or disrupted cells. Disrupted cell preparations offer more information on enzymatic equipment of rickettsiae, because in intact cells the cell membrane is interposed and only a fraction of these enzymes can be placed in contact with exogenous substrates. Most of our knowledge of rickettsial metabolism stems from studies in which purified rickettsial suspensions were used. Such suspensions are capable of oxidizing a number of intermediates of the Krebs cycle. There are minor variations among different rickettsial species in utilizing various substrates. Typhus rickettsiae and R. rickettsi utilize glutamate, glutamine and pyruvate in descending order. Coxiella burneti metabolizes pyruvate as the chief substrate. R. quintana utilizes succinate, alpha-keto-glutarate, glutamine, pyruvate and citrate (Huang, 1967).

It was generally accepted that there is no utilization of glucose by rickettsiae. However, disrupted cells of C. burneti were shown to phosphorylate glucose to glucose-6-phosphate via the hexokinase reaction (Paretski et al., 1962).

The ability of rickettsiae to perform one of the key reactions of the citric acid cycle — the synthesis of citrate — was demonstrated by Paretski et al. (1958).

It is evident that rickettsiae possess a relatively complete energy-yielding apparatus which implies that they have the potential for synthesis of macromolecules in vitro. It seems likely that they contain the enzymatic equipment to synthesize also major cell components from simpler substances. Typhus rickettsiae were shown to incorporate some amino acids from axenic media (Bovarnick and Schneider, 1960; Bovarnick et al., 1959; Fujita et al., 1958; Kohno et al., 1961). Coxiella burneti extracts are also able to incorporate amino acids (MALLAVIA and PARETSKI, 1967). The synthesis of serine and citrulline was also demonstrated (MYERS and PARETSKI, 1961; MALLAVIA and PARETSKI, 1963). The cell-free preparations of Coxiella burneti can incorporate ribonucleoside triphosphates into polyribonucleotides in a reaction mediated by its own DNA-dependent RNA polymerase (JONES and PARETSKI, 1967).

Thus, all available results of biochemical studies on rickettsiae support the concept that rickettsiae are indeed true bacteria which have lost some enzyme systems, but have retained many others (PARETSKI et al., 1958). Despite this statement one basic question still remains to be answered: what causes the dependence of rickettsiae upon the host cell?

Let us consider R. quintana: as shown by HUANG (1967), metabolic reactions of this agent, which multiplies extracellularly, are qualitatively indistinguishable from those occurring in other rickettsiae, multiplying intracellularly. However, there is a difference between R. quintana and typhus rickettsiae in that R. quintana does not require a high K^+/Na^+ ratio for stability and metabolic activity, whereas typhus rickettsiae do. This fact indicates the importance of permeability for large and highly charged molecules for microorganisms adapted to intracellular parasitism (MOULDER, 1962). SMADEL (1963) likewise pointed out that the rickettsial dependence on the host might be a consequence of unusual rickettsial permeability which made rickettsiae highly sensitive to changes in the metabolite pools of their host cells. Studies by MYERS et al. (1967) raise a question regarding the actual permeability of rickettsiae. These authors showed that the freezing and thawing which form part of purification procedures injure the cells so that they lose the capacity to regulate their internal electrolyte concentration. They performed experiments on passive permeability properties of R. mooseri in macerated yolk sacs, using plasmolysis by hypertonic NaCl, KCl or sucrose solutions. In the intact R. mooseri, an osmotically active, functional and structural membrane was found which was passively impermeable for the molecules mentioned. An altered permeability for inorganic electrolytes was detected in rickettsiae which had been stored in the frozen state and subjected to a lengthy purification process. Thus, these findings must be taken into account in considering results of metabolic and enzymatic studies with purified rickettsial suspensions.

Apart from the fact that the intracellular environment probably controls the rickettsial permeability and with it maintains a favourable milieu for metabolic processes, the cell probably supplies some unique essential nutrients, e.g. nucleosides, amino acids or ATP. In this respect, only a few data are as yet available and more information is needed before the prospects of successful cultivation of rickettsiae in axenic media can be properly evaluated.

6. Newer Immunological Findings

The basic body of information about rickettsial immunology was gathered during World War II. After a period of lessened interest in this field, the work was intensified again in the past few years. Many findings were revised and analysed on the basis of modern laboratory approaches. With regard to rickettsial antigens, the main aim has been their isolation, characterization and localization by various physical, biochemical and serological methods.

In this chapter I should like to touch only briefly upon recent results of immunological investigations regarding the typhus group of rickettsiae, as well as Coxiella burneti, because a review of this field has already been published (BREZINA, 1968).

It is beyond any doubt that the cell wall of rickettsiae — as in the case of bacteria — is the main site of their antigenic components. In the typhus group, the first data supporting this view were reported by SCHAECHTER et al. (1957) who studied R. mooseri. FRYGIN and SIWECKA (1965) and FRYGIN (1966) described the chemical and immunological properties of R. prowazeki. In the recent studies of WOOD and WISSEMAN (1967a, b) the cell walls of R. mooseri, obtained by mechanical disruption of cells, were subjected to chemical and immunological analysis. Most of the known immunogenic and antigenic activities of this agent appear to reside in the cell wall. It contains most of the soluble antigen, the erythrocyte-sensitizing substance, the bulk of immunizing activity and skin reacting ability, equal to that of the intact organism, and an endotoxin.

The importance of the soluble antigen of R. prowazeki in conferring immunity has recently been reinvestigated by GOLINEVICH and VORONOVA (1968). They used soluble antigen obtained from rickettsial suspensions by the classical ether method and precipitation by ammonium sulphate. On the basis of its capacity to confer protection, the authors refer to it as the "protective antigen". It is thermolabile, type-specific, and induces complement-fixing, haemagglutinating and toxin-neutralizing antibodies. The relationship of this antigen to that described by BALAYEVA et al. (1966) is not yet clear. BALAYEVA's antigenic substance is not released by ether treatment and its amount is proportional to that of infective rickettsiae. Information on the relationship between antigenic activities and specific structures within the cell wall is needed for characterization of these two kinds of soluble antigen.

Important data were obtained recently regarding the typhus group immunology which provide a basis for serological differentiation between a primary attack of disease and recrudescent typhus (Brill-Zinsser disease). It is generally accepted that 19 S antibodies are associated with the primary response, whereas the 7 S antibodies appear later. Aside from various physical and chemical methods for differentiation of 19 S and 7 S antibodies, it is possible to distinguish between them by treatment of sera with the sulfhydryl compound, ethanethiol, and by their inactivation at 60° C (MURRAY et al., 1965a). Further, primary epidemic typhus sera require high doses of antigen, a phenomenon described by MURRAY et al. (1965b). Finally, differences in the immunoelectrophoretic pattern as well as in the inactivating effect of 2-mercaptoethanol were found between primary epidemic typhus sera and Brill-Zinsser disease sera (MURRAY et al., 1965c). VORONOVA (1968) described a very simple method for differentiating the 19 S and 7 S antibodies by a passive haemagglutination test in the presence of cystein. She studied also the sequence of formation of antibodies with different electrophoretic mobility in experimentally infected rabbits. The results supported the possibility of a serological differentiation between primary and repeated R. prowazeki infections (VORONOVA, 1966).

Study of C. burneti antigens stemmed from the discovery of the phase variation phenomenon (STOKER and FISET, 1956). There are two main antigens in this agent, one of them being obligatory (antigen II) and the second one being facultative, appearing or disappearing under certain circumstances (antigen I). Phase I organisms possess both antigens, those in phase II reveal only antigen II. From phase I organisms, a protective antigen can be extracted by trichloroacetic acid (BREZINA and URVÖLGYI, 1961, 1962; ANACKER et al., 1962), dimethylsulfoxide or dimethylacetamide (ORMSBEE et al., 1962). The trichloroacetic acid extract can be converted by phenol treatment to a hapten by removing the protein component from the lipid — protein-polysaccharide complex (BREZINA et al., 1962; ANACKER et al., 1963). The trichloroacetic acid extract displays a pyrogenic effect in guinea pigs (BREZINA et al., 1965). As shown by chromatographic analysis on DEAE cellulose columns, the crude TCA extract exhibits a marked heterogeneity (BREZINA and SCHRAMEK, 1968).

The presence of antigen I in C. burneti affects the properties of this agent in many respects. There are many points of resemblance between the rough-smooth phase variation of Diplococcus pneumoniae and the phase variation of C. burneti. Phase II organisms tend to agglutinate spontaneously and to be agglutinable by normal sera, whereas phase I organisms resist phagocytosis unless sensitized by specific antibodies (BREZINA and KAZÁR, 1963, 1965; WISSEMAN et al., 1967). These facts may contribute to a higher virulence of phase I Coxiellae. Phase I organisms can be detected by means of the fluorescent antibody technique only with sera containing antibodies against antigen I (BREZINA and KOVÁČOVÁ, 1966).

To date very little is known about the nature of antigenic phase II component. It is possible to extract part of this antigen from purified phase II organisms by means of urea (BREZINA, 1968).

Highly purified and chemically well defined antigens are necessary for demonstration of a particular kind of induced antibodies. In Q fever, various antibodies can be detected by a number of serological methods. It has not yet been determined what type of antibody actually reflects the state of immunity. It is merely known that they are localized in the 7 S slow gamma-globulin fraction of immune serum along with the agglutinins and complement fixing antibodies as shown for R. prowazeki and C. burneti by ORMSBEE et al. (1968).

7. Latency and Persistence of Rickettsiae

A well-known attribute of rickettsiae is their long term persistence in natural animal hosts, in man, and in experimentally infected animals. In arthropods, rickettsial infections often persist for the life of the host and may also be transmitted transovarially to further generations.

Little is known about the nature and mechanism of the persistent infections. There are probably many factors, determined by both rickettsiae and host, which are involved in this phenomenon. The questions which arise with regard to rickettsial persistence may be listed as follows (MARMION, 1959): (1) in what state do the rickettsiae persist; (2) where do the rickettsiae persist; (3) what are the factors which permit survival of rickettsiae under the attack of the host's defence mechanisms, and (4) what are the factors which restrict rickettsial multiplication or lead to increased multiplication and change from the latent state to recrudescent illness.

As for antibodies, it seems that they have little effect on rickettsiae within cells. Antibiotics likewise are unable to free the host organism of rickettsiae despite the fact that tissue cultures can be cured (SMADEL, 1963). In our experiments (BREZINA and KORDOVÁ, 1957) chlortetracycline a terramycin did not reduce the duration of persisting C. burneti infections of mice. The early and protracted treatment of volunteers infected with R. tsutsugamushi with chloramphenicol resulted in a carrier state which, after the treatment was stopped and in the absence of antibodies produced clinical disease (SMADEL, 1963).

In would be paradoxical to suppose that rickettsiae might reveal a mode of persistence similar to intracellular survival of genetic material of viruses. However, in the case of C. burneti, there is the possibility that filterable particles may persist (KORDOVÁ, 1959a, 1960b). Another explanation might be that rickettsiae persist in an incomplete form as suggested by PRICE et al. (1958) in regard to persistence of R. prowazeki in man which may result in Brill-Zinsser disease.

The clue to the elucidation of rickettsial persistence may lie in the rickettsia-host cell relationship studied on the cellular level. So far very little work has been done in this field.

MINIMASHIMA and MORI (1964) and MINIMASHIMA (1965) studied the persistence of R. sennetsu in FL cell cultures. They observed the persistence of the agent for 20 passages over a total period of 504 days. In all passages the presence of rickettsiae was confirmed by microscopic examination and by titration in mice. As a result of continuous association with FL cells, R. sennetsu became attenuated. Results differing from those of the Japanese authors were reported by POSPÍŠIL (1966) who studied the persistent infection of monkey kidney cell cultures (stable line OL) with C. burneti. A total of 14 passages was carried out within a period of 8 months. The microscopic examination of stained preparations as well as immunofluorescence technique revealed Coxiellae only in the first 4—5 passages. In later passages, no manifestly infected cells were found, but the presence of the agent could be detected by inoculation of yolk sacs with cell suspensions. In further experiments (POSPÍŠIL, 1967) 5th passage cells were cloned. After five reclonings C. burneti was recovered in cells following inoculation of yolk sacs. It is not clear in what form C. burneti persists in OL cells. Whether filterable particles are involved in the mechanism of persistence remains to be answered.

The Japanese authors and Pospíšil observed cells undergoing division and both daughter cells contained rickettsiae. This fact must be taken into account as one of the possible mechanisms of maintenance of persistent infection as well as cell-to-cell transmission via intracellular bridges.

Finally, the observations concerning the so called "resting" forms of rickettsiae (Kokorin, 1968) should be recalled; these are able to persist in cells for a very long time and could represent one of the mechanisms by which latent rickettsial infections are maintained.

8. Rickettsiae and Interferon

In the past 3—4 years, some attention was paid to the possibility that rickettsiae might induce interferon or interferon-like substances. First observations were made by Hopps et al. (1964) who found in the tissue culture fluid of chick embryo cells infected with Karp strain of R. tsutsugamushi a substance which inhibited virus multiplication. As challenge viruses Sindbis, vesicular stomatis virus and the PR 8 strain of influenza were employed. The inhibitory substance possessed many of the physical and biological properties of interferon. Its molecular weight was found to be similar to virus-induced interferon produced by the same type of cells (Hopps and Merigan, unpublished data, cited by Merigan, 1967).

As for the induction of interferon in animals, Kazár (1966) first demonstrated an interferon-like inhibitor in mouse sera following intravenous administration of partially purified and concentrated suspensions of C. burneti or R. prowazeki from infected yolk sacs. The inhibitory activity was tested in L cells against encephalomyocarditis (EMC) virus and was highest in sera taken 3—5 hours after inoculation. Preliminary tests indicated that this substance conformed to interferon according to the usual criteria.

The report of Hopps et al. (1964) is thus far the only one on the ability of rickettsiae to induce interferon synthesis in tissue cultures. There are additional data on interference between rickettsiae and viruses in which the role of interferon has not been clarified. Thus, Tribble et al. (1965) demonstrated interference between R. rickettsi and Venezuelan equine encephalomyelitis virus in L cell cultures. However, no inhibitory activity was found in tissue culture fluid. The authors suggest that under certain conditions rickettsiae may influence the internal milieu of cells in such a way that viruses cannot multiply in them. Further, Kazár (1967) demonstrated that infection of chick embryo fibroblast cultures with C. burneti interfered with multiplication of several cythopathic viruses. Interference was pronounced with Sindbis, Western equine encephalomyelitis and vesicular stomatitis viruses and less striking with pseudorabies and vaccinia viruses. No interference was observed between C. burneti and Newcastle disease virus. Interference depended on multiplication of C. burneti, since a dose of chlortetracycline which prevented multiplication of rickettsia also prevented the establishment of interference. The mechanism of this kind of interference is not clear.

So far no inhibitory activity of virus-induced or rickettsia-induced interferons on the multiplication of rickettsiae has been demonstrated. In contrast, Chlamydiae which, like rickettsiae, resemble bacteria rather than viruses, are inhibited by virus-induced interferon in their multiplication in tissue cultures as observed by HANNA et al. (1966) with the TRIC agent LB 1. REINICKE et al. (1967) likewise demonstrated the inhibition of a TRIC agent (Bour strain) in ovo by high doses of purified interferon induced by influenza virus B (Lee).

The activity of interferons is generally assayed against riruses and the question arises how might one explain the effect of interferon on intracellular microorganisms other than viruses. According to HANNA et al. (1966), the inhibition of TRIC agent replication (and probably intracellular bacteria) may be explained by the basic similarity of all biosynthetic patterns. FRESHMAN et al. (1966) called attention to ribosomes and the part they play in the activity of interferon. So far it is not known whether both Chlamydiae and rickettsiae employ their own ribosomes or those of their host cells. In any case, further experimental work is needed to elucidate the mechanism by which interferon affects intracellular microorganisms such as Chlamydiae and perhaps also rickettsiae.

References

ALLISON, A. C., and H. R. PERKINS: Presence of the cell walls like those of bacteria in rickettsiae. Nature (Lond.) **180**, 796—798 (1960).

ANACKER, R. L., K. FUKUSHI, E. G. PICKENS, and D. B. LACKMAN: Electron microscopic observation of the development of Coxiella burnetii in the chick yolk sac. J. Bact. **88**, 1130—1138 (1964).

— W. T. HASKINS, D. B. LACKMAN, E. RIBI, and E. G. PICKENS: Conversion of the phase I antigen of Coxiella burnetii to hapten by phenol treatment. J. Bact. **85**, 1165—1170 (1963).

— D. B. LACKMAN, E. G. PICKENS, and E. G. RIBI: Antigenic and skin-reactive properties of fractions of Coxiella burnetii. J. Immunol. **89**, 145—153 (1962).

— E. G. PICKENS, and D. B. LACKMAN: Details of the ultrastructure of Rickettsia prowazekii grown in the yolk sac. J. Bact. **94**, 260—262 (1967).

ANDERSON, D. R., E. H. HOPPS, M. F. BARILE, and B. C. BERNHEIM: Comparison of the ultrastructure of several rickettsiae, ornithosis virus and Mycoplasma in tissue cultures. J. Bact. **90**, 1387—1404 (1965).

BALAYEVA, N. M., L. P. ZUBOK, and V. N. NIKOLSKAYA: Soluble antigen of Rickettsia prowazeki. Acta virol. **10**, 161—168 (1966).

BELL, E. J., G. M. KOHLS, H. G. STOENNER, and D. B. LACKMAN: Nonpathogenic rickettsias related to the spotted fever group isolated from ticks Dermacentor variabilis and Dermacentor andersoni from eastern Montana. J. Immunol. **90**, 770—781 (1963).

BIRD, R. G., N. KORDOVÁ, and J. ŘEHÁČEK: Fine structure of Rickettsia prowazeki in the haemocytes of ticks Hyalomma dromedarii. Acta virol. **11**, 60—62 (1967).

BOVARNICK, M. R., and L. SCHNEIDER: The incorporation of glycine-1-C^{14} by typhus rickettsiae. J. biol. Chem. **235**, 1727—1731 (1960).

—, and H. WALTER: The incorporation of labelled methionine by typhus rickettsiae. Biochim. biophys. Acta (Amst.) **33**, 414—422 (1959).

BOZEMAN, M., J. W. HUMPHRIES, and J. M. CAMPBELL: A new group of rickettsia-like agents recovered from guinea pigs. Acta virol. **12**, 87—93 (1968).

BREZINA, R.: Antigenic components of Coxiella burneti in relation to some properties of the agent. Proc. of the 7th Int. Congr. on trop. Med. Malaria 3, 262—265 (1964).
— New advances in the study of rickettsial antigens. Zbl. Bakt., I. Abt. Orig. **206**, 313—321 (1968).
—, and J. KAZÁR: Phagocytosis of Coxiella burneti and the phase variation phenomenon. Acta virol. **7**, 746 (1963).
— — Study of the antigenic structure of Coxiella burneti. IV. Phagocytosis and opsonization in relation to the phases of C. burneti. Acta virol. **9**, 268—274 (1965).
—, and N. KORDOVÁ: Príspevok k účinku aureomycínu a teramycínu na experimentálnu infekciu myšiek C. burneti. Vet. Čas. **6**, 184—191 (1957).
—, and E. KOVÁČOVÁ: Study of the antigenic structure of Coxiella burneti. V. Detection of phase I of Coxiella burneti by indirect fluorescent antibody technique. Acta virol. **10**, 462—466 (1966).
—, and Š. SCHRAMEK: Study of the antigenic structure of Coxiella burneti. VI. Chromatographic analysis of phase I antigen of C. burneti on DEAE-cellulose. Acta virol. **12**, 68—72 (1968).
— —, and J. URVÖLGYI: Study of the antigenic structure of Coxiella burneti. II. Purification of phase I antigenic component obtained by means of trichloracetic acid. Acta virol. **6**, 278—279 (1962).
— — — Study of the antigenic structure of Coxiella burneti. III. Pyrogenic effect of phase I antigen in experimental guinea pigs. Acta virol. **9**, 180—185 (1965).
—, and J. URVÖLGYI: Extraction of Coxiella burneti phase I antigen by means of trichloracetic acid. Acta virol. **5**, 193 (1961).
— — Study of the antigenic structure of Coxiella burneti. I. Extraction of phase I antigen by means of trichloracetic acid. Acta virol. **6**, 84—88 (1962).
BURGDORFER, W.: Observation on Rickettsia canada, a new member of the typhus group of rickettsiae. J. Hyg. Epidem. (Praha) **12**, 26—31 (1968).
—, and R. A. ORMSBEE: Development of Rickettsia prowazeki in certain species of Ixodid ticks. Acta virol. **12**, 36—40 (1968).
—, and M. G. R. VARMA: Trans-stadial and transovarial development of disease agents in arthropods. Annual Rev. Entomol. **12**, 347—376 (1967).
BURNET, F. M.: Principles of animal virology. New York: Academic Press Inc. 1955.
DAVIS, G. E., and H. E. COX: A filter-passing infectious agent isolated from ticks. I. Isolation from Dermacentor andersoni, reactions in animal and filtration experiments. Publ. Hlth Rep. (Wash.) **53**, 2259—2267 (1938).
FISET, P.: Phase variation of Rickettsia (C.) burnetii. Study of the antibody response in guinea pigs and rabbits. Canad. J. Microbiol. **3**, 435—445 (1957).
—, and R. A. ORMSBEE: The antibody response to antigens of Coxiella burneti. Zbl. Bakt., I. Abt. Orig. **206**, 321—329 (1968).
FRESHMAN, M. M., T. G. MERIGAN, J. S. REMINGTON, and I. E. BROWNLEE: In vitro and in vivo antiviral action of an interferon-like substance induced by Toxoplasma gondii. Proc. Soc. exp. Biol. (N.Y.) **123**, 862—866 (1966).
FRYGIN, C.: Wlaściwości immunologiczne ściany komórkowej Rickettsia prowazeki. Med. dośw. Mikrobiol. **18**, 117—125 (1966).
—, and M. SIWECKA: Otrzymywanie preparatow ściany komórkowej Rickettsia prowazeki a ich wstepna analiza. Med. dośw. Mikrobiol. **17**, 143—152 (1965).
FUJITA, K., S. KOHNO, and A. SHISHIDO: The incorporation of methionine by typhus rickettsiae. Jap. J. med. Sci. Biol. **12**, 387—390 (1958).
FUKUDA, T., Y. KEIDA, and T. KITAO: Studies on causative agent of "Hyuga netsu" disease. Med. and Biol. **23**, 200—205 (1954) (in Japanese).
GIMENEZ, D. F.: Gram staining of Coxiella burnetii. J. Bact. **90**, 834—835 (1965).
GOLINEVICH, E. M., and Z. A. VORONOVA: Poverkhnostnii protektivnii antigen rikketsii Provačeka. J. Hyg. Epidem. (Praha) **12**, 23—30 (1968).

Handley, J., D. Paretsky, and J. Stueckmann: Electron microscopic observations of Coxiella burneti in the guinea pig. J. Bact. **94**, 263—267 (1967).

Hanna, L., T. C. Merigan, and J. Jawetz: The inhibition of TRIC agents by virus-induced interferon. Proc. Soc. exp. Biol. (N.Y.) **122**, 417—421 (1966).

Higashi, N.: Recent advances in electron microscope studies on ultrastructure of rickettsiae. Zbl. Bakt., I. Abt. Orig. **206**, 277—283 (1968).

Hoogstraal, H.: Ticks in relation to human diseases caused by Rickettsia species. Ann. Rev. Entomol. **12**, 377—420 (1967).

Hopps, H. E., S. Kohno, M. Kohno, and J. E. Smadel: Production of interferon in tissue cultures infected with Rickettsia tsutsugamushi. Bact. Proc. 115—116 (1964).

Huang, K.: Metabolic activity of the Trench fever rickettsia, Rickettsia quintana. J. Bact. **93**, 853—859 (1967).

Ito, S., and J. W. Vinson: Fine structure of Rickettsia quintana cultivated in vitro and in the louse. J. Bact. **89**, 481—495 (1965).

Jones, F., and D. Paretski: Physiology of rickettsiae. VI. Host-independent synthesis of polyribonucleotides by Coxiella burneti. J. Bact. **93**, 1063—1068 (1967).

Kazár, J.: Interferon-like inhibitor in mouse sera induced by rickettsiae. Acta virol. **10**, 277 (1966).

— Interference between Coxiella burneti and some cytopathic viruses in chick embryo cell cultures. (To be published) (Paper presented at Int. Symp. on Rickett., Smolenice near Bratislava, September 1967).

Kohno, S., Y. Natsume, and A. Shishido: Studies on metabolism of rickettsiae. II. Incorporation of C^{14} amino acid mixture and S^{35} methionine by Rickettsia mooseri. Jap. J. med. Sci. Biol. **14**, 213—232 (1961).

Kokorin, I. N.: Biological peculiarities of the development of rickettsiae. Acta virol. **12**, 31—35 (1968).

Kordová, N.: Niektoré výsledky štúdia Coxiella burneti v tkanivových kulturach. Čs. Mikrobiol. **3**, 220—224 (1958).

— Filterable particles of Coxiella burneti. Acta virol. **3**, 25—36 (1959a).

— Microscopic study of tissue cultures infected with filterable particles of Coxiella burneti. Folia microbiol. (Praha) **4**, 237—239 (1959b).

— Study of antigenicity and immunogenicity of filterable particles of Coxiella burneti. Acta virol. **4**, 56—62 (1960a).

— Latent infections in animals by filterable particles of Coxiella burneti. Acta virol. **4**, 173—183 (1960b).

— Die Vermehrung der Rickettsia prowazeki in L-Zellen. I. Lichtmikroskopische Untersuchungen. Arch. ges. Virusforsch. **15**, 698—706 (1965).

—, and R. Brezina: Multiplication dynamics of phase I and II Coxiella burneti in different cell cultures. Acta virol. **7**, 84—87 (1963).

—, and E. Kováčová: Replication of Rickettsia prowazeki in L-cells as revealed by immunofluorescence. Acta virol. **11**, 252—255 (1967).

— — Histochemical and fluorescent antibody studies on the early stages of infection of L cells with Coxiella burneti. Acta virol. **12**, 23—30 (1968).

—, and P. Kvíčala: Coxiella burneti in tissue cultures, studied by the optic microscope and in phase contrast. Folia microbiol. (Praha) **7**, 89—92 (1962).

—, and J. Řeháček: Experimental infection of ticks in vivo and their organs in vitro with filterable particles of Coxiella burneti. Acta virol. **3**, 201—209 (1959a).

— — Cultivation of ultrafilterable particles of Coxiella burneti in tick organism. Folia microbiol. (Praha) **4**, 275—279 (1959b).

— — Microscopic examination of the organs of ticks infected with Rickettsia prowazeki. Acta virol. **8**, 465—469 (1964).

— M. Rosenberg u. E. Mrena: Die Vermehrung der Rickettsia prowazeki in L-Zellen. II. Elektronenmikroskopische Untersuchungen an infizierten Gewebezellen in Dünnschnitten. Arch. ges. Virusforsch. **15**, 708—720 (1965).

KOVÁČOVÁ, E., u. N. KORDOVÁ: Die Vermehrung der Rickettsia prowazeki in L-Zellen. III. Acridin-Orange-Fluoreszenz und Ultraviolett-Mikroskopie. Arch. ges. Virusforsch. **19**, 57—62 (1966).

LACKMAN, D. B., E. J. BELL, H. G. STOENNER, and E. G. PICKENS: The Rocky Mountains spotted fever group of rickettsias. Hlth Lab. Sci. **2**, 135—141 (1965).

— R. R. PARKER, and R. K. GERLOFF: Serological characteristics of a pathogenic rickettsia occurring in Amblyomma maculatum. Publ. Hlth Rep. (Wash.) **64**, 1342—1349 (1949).

MALLAVIA, L., and D. PARETSKI: Studies on the physiology of rickettsiae. V. Metabolism of carbamyl phosphate by Coxiella burneti. J. Bact. **86**, 232—238 (1963).

— — Physiology of Rickettsiae. VII. Amino acid incorporation by Coxiella burneti and by infected hosts. J. Bact. **93**, 1479—1483 (1967).

MARMION, B. P.: Latency and rickettsial infections. Proc. 6th Int. Congr. trop. Med. Malaria **5**, 711—717 (1959).

MCKIEL, J. A., E. J. BELL, and D. B. LACKMAN: Rickettsia canada: A new member of the typhus group of rickettsiae isolated from Haemaphysalis leporis palustris ticks in Canada. Canad. J. Microbiol. **13**, 503—510 (1967).

MERIGAN, T. C.: Various molecular species of interferon induced by viral and non-viral agents. Bact. Rev. **31**, 138—144 (1967).

MINIMASHIMA, Y.: Persistent infection of Rickettsia sennetsu in cell culture system. Jap. J. Microbiol. **9**, 75—86 (1965).

—, and R. MORI: Persistent infection of FL cells by Rickettsia sennetsu (S. todai). J. Bact. **88**, 1195—1196 (1964).

MISAO, T., and Y. KOBAYASHI: Studies on infectious mononucleosis. I. On the isolation of causative agent from blood, bone marrow and lymph gland with mice. Tokyo Med. J. **71**, 683—686 (1954) (in Japanese).

MOULDER, J. W.: The biochemistry of intracellular parasitism. Chicago: Chicago University Press 1962.

MURRAY, E. S., J. A. GAON, J. M. O'CONNOR, and M. MULAHASANOVIĆ: Serologic studies of primary epidemic and recrudescent typhus (Brill-Zinsser disease). I. Differences in complement fixing antibodies: high antigen requirement and heat lability. J. Immunol. **94**, 723—733 (1965b).

— J. M. O'CONNOR, and J. A. GAON: Differentiation of 19 S and 7 S complement fixing antibodies in primary versus recrudescent typhus by either ethanethiol or heat. Proc. Soc. exp. Biol. (N.Y.) **119**, 291—297 (1965a).

— — — Serologic studies of primary epidemic and recrudescent typhus (Brill-Zinsser disease). Differences in immunoelectrophoretic patterns, response to 2-mercaptoethanol and relationships to 19 S and 7 S antibodies. J. Immunol. **94**, 734—740 (1965c).

MYERS, W. F., and D. PARETSKI: Synthesis of serine by Coxiella burneti. J. Bact. **82**, 761—763 (1961).

— P. J. PROVOST, and C. L. WISSEMAN, JR.: Permeability properties of Rickettsia mooseri. J. Bact. **93**, 950—960 (1967).

NERMUT, M., Š. SCHRAMEK, and R. BREZINA: Electron microscopy of Coxiella burneti phase I and II. (To be published.) (Paper presented at Int. Symp. on Rickett., Smolenice near Bratislava, September 1967.)

ORMSBEE, R. A.: The growth of Coxiella burnetii in embryonated eggs. J. Bact. **63**, 73—86 (1952).

— E. J. BELL, and D. B. LACKMAN: Antigens of Coxiella burneti. I. Extraction of antigens with nonaqueous organic solvents. J. Immunol. **88**, 741—749 (1962).

— M. PEACOCK, G. TALLENT, and J. J. MUNOZ: An analysis of the immune response to rickettsial antigens in guinea pigs. Acta virol. **12**, 78—82 (1968).

PARETSKI, D.: Biochemistry of rickettsiae and their infected hosts, with special reference to Coxiella burnetii. Zbl. Bakt., I. Abt. Orig. **206**, 283—291 (1968).

Paretski, D., R. A. Consigli, and C. M. Downs: Studies on the physiology of rickettsiae. III. Glucose phosphorylation and hexokinase activity in Coxiella burnetii. J. Bact. **83**, 538—543 (1962).

— C. M. Downs, R. A. Consigli, and B. K. Joyce: Studies on the physiology of rickettsiae. I. Some enzyme systems of Coxiella burnetii. J. infect. Dis. **103**, 6—11 (1958).

Parker, R. R., G. M. Kohls, G. W. Cox, and G. E. Davis: Observations on an infectious agent from Amblyomma maculatum. Publ. Hlth Rep. (Wash.) **54**, 1482—1484 (1939).

Philip, C. B.: A review of growing evidence that domestic animals may be involved in cycles of rickettsial zoonoses. Zbl. Bakt., I. Abt. Orig. **206**, 343—353 (1968).

Pospíšil, V. F.: Persistent infection of cell cultures with Coxiella burneti. Acta virol. **10**, 542—548 (1966).

— Biological changes of Coxiella burneti in the course of its persistence in monkey kidney (OL) cells. (To be published.) (Paper presented at Int. Symp. on Rickett., Smolenice near Bratislava, September 1967).

Price, W. H., H. Emerson, H. Nagel, R. Blumberg, and S. Talmadge: Ecologic studies on the interepidemic survival of louse-borne epidemic typhus fever. Amer. J. Hyg. **67**, 154—178 (1958).

Řeháček, J.: Development of animal viruses and rickettsiae in ticks and mites. Ann. Rev. Entomol. **12**, 1—24 (1965).

Reinicke, V., C. H. Mordhorst, and E. Schonne: In vivo inhibition of the growth of TRIC agents by purified species-specific interferon. Acta path. microbiol. scand. **69**, 478—479 (1967).

Reiss-Gutfreund, R. J.: Un nouveau réservoir de virus pour Rickettsia prowazeki: les animaux domestiques et leur tiques. Bull. Path. Exot. **49**, 946—1021 (1956).

Roberts, A. N., and C. M. Downs: Study on the growth of Coxiella burneti in the L strain mouse fibroblasts and chick embryo cells. J. Bact. **77**, 194—204 (1959).

Rosenberg, M., and N. Kordová: Study of intracellular forms of Coxiella burneti in the electron microscope. Acta virol. **4**, 52—55 (1960).

— — Multiplication of Coxiella burneti in Detroit-6 cell cultures. An electron microscope study. Acta virol. **6**, 176—180 (1962).

Schaechter, M., F. M. Bozeman, and J. E. Smadel: Study on the growth of rickettsiae. II. Morphologic observations of living rickettsiae in tissue culture cells. Virology **3**, 160—171 (1957).

Schaechter, M. A., A. J. Tousimis, Z. A. Cohn, H. Rosen, J. Campbell, and F. E. Hahn: Morphological, chemical and serological studies of the cell walls of Rickettsia mooseri. J. Bact. **74**, 822—829 (1957).

Shishido, A., S. Honjo, M. Suganuma, S. Ohtaki, M. Hikita, T. Fujiwara, and M. Takasaka: Studies on infectious mononucleosis induced in the monkey by experimental infection with Rickettsia sennetsu. I. Clinical observations and etiological investigations. Jap. J. med. Sci. **18**, 73—83 (1965).

Shkolnik, L. Ya., B. G. Zatulovski, and N. M. Shestopalova: Ultrastructure of Rickettsia prowazeki. An electron microscope study of ultrathin sections from infected louse guts and chick embryo yolk sac. Acta virol. **10**, 260—265 (1966).

Smadel, J. E.: Intracellular infection and the carrier state. Science **140**, 153—160 (1963).

Stoker, M. G. P.: Q fever in Great Britain. The causative agent. Lancet **1950 II**, 616—620.

—, and P. Fiset: Phase variation of the Nine Mile and other strains of Rickettsia burneti. Canad. J. Microbiol. **2**, 310—321 (1956).

Tribble, H. R., H. B. Rees, and H. J. Hearn: Interaction between Venezuelan equine encephalomyelitis virus and Rickettsia rickettsii in suspended L cells. Bact. Proc. 98 (1965).

Urbach, H., W. Linss, and A. Stelzner: Morphological changes in cells of guinea pig testes after infection with Coxiella burneti. Acta virol. **12**, 11—14 (1968).

Vinson, J. W.: In vitro cultivation of the rickettsial agent of Trench fever. Bull. Wld Hlth Org. **35**, 155—164 (1966).

Voronova, Z. A.: Raspredelenie antitel v belkovych frakciach syvorotok posle pervichnogo i povtornogo zarazhenia krolikov rikketsiami Provacheka. Zh. Mikrobiol. (Mosk.) **3**, 134—138 (1966).

— Differentiation of 19 S and 7 S rickettsial antibodies by passive haemagglutination reaction with cysteine. Acta virol. **12**, 73—77 (1968).

Weiss, E.: Comparative metabolism of rickettsiae and other host dependent bacteria. Zbl. Bakt., I. Abt. Orig. **206**, 292—298 (1968).

Weyer, F.: Extracellular development of rickettsiae. Acta virol. **12**, 44—48 (1968).

Wisseman, C. L., Jr.: Some biological properties of rickettsiae pathogenic for man. Zbl. Bakt., I. Abt. Orig. **206**, 299—313 (1968).

— P. Fiset, and R. A. Ormsbee: Interaction of rickettsiae and phagocytic host cells. V. Phagocytic and opsonic interactions of phase I and phase II Coxiella burneti with normal and immune human leukocytes and antibodies. J. Immunol. **99**, 669—674 (1967).

Wood, W. H., and C. L. Wisseman, Jr.: The cell wall of Rickettsia mooseri. I. Morphology and chemical composition. J. Bact. **93**, 1113—1118 (1967a).

— — Studies of Rickettsia mooseri cell walls. II. Immunologic properties. J. Immunol. **98**, 1224—1230 (1967b).

Zdrodovski, P. F., and E. H. Golinevich: The rickettsial diseases. Oxford-London-New York-Paris: Pergamon Press 1960.

Zubok, L. O.: Fiziologia i biochimia rikketsij. Usp. sovrem Biochim. **61**, 426—442 (1966).

Max-Planck-Institut für Molekulare Genetik Berlin (Dahlem),
Ehrenbergstr. 26—28

Die molekularen Grundlagen der Photoreaktivierung

HEINZ SCHUSTER

Mit 7 Abbildungen

Inhalt

I. Einleitung

In den Desoxyribonucleinsäuren (DNS)[1] der Zellen und der Viren ist die genetische Information enthalten, welche die identische Reproduktion der biologischen Systeme gewährleistet. Neben der DNS kommt bei einer Reihe von Viren auch noch Ribonucleinsäure (RNS) als Informationsträger vor. Kommt es in der natürlichen Umgebung oder durch künstlichen Eingriff zu einer Veränderung in der chemischen Struktur des Nucleinsäuremoleküls, so kann dies verschiedene Konsequenzen haben:

[1] Die folgenden Abkürzungen werden verwendet: DNS = Desoxyribonucleinsäure, RNS = Ribonucleinsäure, HCR = Wirtszellreaktivierung („host cell reactivation"), PR = Photoreaktivierung, UV = Ultraviolett(-Licht), BU = Bromuracil; T=T, T=C, C=C, T=U, U=U = dimere Verbindungen von Thymin mit Thymin, Thymin mit Cytosin, Cytosin mit Cytosin, Thymin mit Uracil und Uracil mit Uracil.

1. Die veränderte Struktur der Nucleinsäure hat keinen Einfluß auf die Fähigkeit des Systems, sich selbst zu reproduzieren. Die für den Reproduktionsvorgang verantwortlichen Enzyme sind nicht in der Lage, bestimmte Anomalien in der Struktur zu erkennen. Die Nachkommen enthalten aber wieder die ursprüngliche Nucleinsäurestruktur [z.B. DNS in Escherichia coli und Tabakmosaikvirus RNS, die primär Bromuracil (BU) anstelle von Thymin (ZAMENHOF et al., 1956) bzw. Fluoruracil anstelle von Uracil (GORDON und STAEHELIN, 1959) enthalten].

2. Die Strukturumwandlung führt zu einem teilweise neuen Informationsgehalt (Mutation). Das System behält die Fähigkeit zur Selbstreproduktion, eine oder mehrere seiner Funktionen sind aber irreversibel verändert worden. Das Nucleinsäuremolekül behält die veränderte Struktur bei, wobei die primäre Veränderung sich direkt oder auch erst in einer sekundären Umwandlung permanent manifestiert.

3. Die Eingriffe haben das Nucleinsäuremolekül so verändert, daß das System nicht mehr in der Lage ist, lebenswichtige Funktionen auszuüben; die Veränderung wirkt somit letal.

Neben dieser rein passiven Begegnung der Veränderung im genetischen Apparat besitzen Organismen aber auch die Fähigkeit, der letalen (oder auch mutagenen) Wirkung einer Veränderung *aktiv durch Eliminierung des Schadens* zu begegnen. Es kommt dabei zu einer *Reaktivierung* des biologischen Systems. *Photoreaktivierung* und *Wirtszellreaktivierung* ("host cell reactivation", HCR), zwei in ihrer Wirkungsweise verschiedene, molekulare Reparaturprozesse, bewirken diese Regeneration. Beide Arten der Reaktivierung sind genetisch bestimmt, d.h. sie werden vererbt und können durch Mutation verändert werden bzw. verlorengehen.

Bei der Untersuchung der Wirkungsweise von ultravioletten Lichtstrahlen (UV) auf Conidien von Streptomyces griseus hatte KELNER (1949) gefunden, daß die letale Wirkung auf die Zellen von den Bedingungen abhängt, unter denen die Zellen gezüchtet werden. Conidien, die nach der UV-Bestrahlung im sichtbaren Licht gehalten wurden, zeigten (per Dosis) eine weitaus höhere Chance zu überleben wie die zum Vergleich im Dunkeln aufbewahrten Zellen. Dieselbe Beobachtung machte DULBECCO (1949) bei UV-inaktivierten T2-Bakteriophagen, wenn die mit den Phagen infizierten *E. coli* Zellen im Licht statt im Dunkeln aufbewahrt wurden. Dieser als *Photoreaktivierung* (PR) bezeichnete Effekt ist ein enzymatischer Reaktivierungsmechanismus, bei dem ein Enzym der Zelle unter Ausnutzung der Energie des sichtbaren Lichts ganz spezifisch eine bestimmte Art der Veränderung in der DNS wieder rückgängig macht. Die hochmolekulare Struktur der Nucleinsäure bleibt während des enzymatischen Reparaturprozesses erhalten. Das photoreaktivierende Enzym unterscheidet nicht zwischen UV-Licht-geschädigter zelleigener oder Virus-DNS.

Im Gegensatz zur Photoreaktivierung ist die *Wirtszellreaktivierung* (HCR) (GAREN und ZINDER, 1955) ein komplizierterer, von der Lichtenergie unabhängiger Reparaturprozeß, an dem wahrscheinlich mehrere Enzyme der Zelle beteiligt sind (HOWARD-FLANDERS und BOYCE, 1966). Diese haben offensichtlich eine geringe Spezifität in bezug auf die Strukturänderung, die in der DNS aufgetreten ist, d.h. sie können viele in ihrer Struktur verschiedenartige Schäden in der DNS eliminieren (HANAWALT und HAYNES, 1965; HOWARD-FLANDERS und BOYCE, 1966). Der veränderte Nucleinsäurebestandteil wird dabei durch eine örtlich und zeitlich begrenzte Auflösung der hochmolekularen Nucleinsäurestruktur als niedermolekulares Bruchstück entfernt (R. B. SETLOW und CARRIER, 1964; BOYCE und HOWARD-FLANDERS, 1964).

Kürzlich ist gezeigt worden, daß Bakterienzellen, die durch Mutation die Fähigkeit zur Wirtszellreaktivierung verloren haben, trotzdem noch eine beträchtliche Zahl künstlich induzierter Schäden in ihrer DNS eliminieren können (RUPP und HOWARD-FLANDERS, 1968). Bei diesem Prozeß, der wie die HCR auch als Dunkelreaktion abläuft, ist die Reparatur abhängig von dem Vorhandensein eines Mechanismus, der Stückaustausch (Rekombination) zwischen verschiedenen DNS-Molekülen bewerkstelligen kann. Fehlen einer Zelle die Fähigkeiten zur Wirtszellreaktivierung *und* zur Reaktivierung durch Rekombination gleichzeitig (Doppelmutante), dann hat ein einziger in der Zell-DNS auftretender Schaden den Zelltod zur Folge (HOWARD-FLANDERS und BOYCE, 1966).

In diesem Beitrag soll über die Photoreaktivierung bei Mikroorganismen und Viren berichtet werden. Bei der Betrachtung soll die Beschreibung der molekularen Vorgänge der Photoreaktivierung, soweit sie uns heute schon bekannt sind, im Vordergrund stehen. In den letzten Jahren sind bereits eine Reihe von zusammenfassenden Beiträgen erschienen, die sich ebenfalls mit dem Phänomen der Photoreaktivierung befassen (JAGGER, 1958; RUPERT, 1964a; RUPERT und HARM, 1966; J. K. SETLOW, 1966a, b).

II. Photochemie von Nucleinsäuren

Die in der Einleitung erwähnten Reaktivierungsphänomene sind alle bei Bestrahlung von Mikroorganismen und Viren mit UV-Licht zum ersten Male beobachtet worden. Wenn sich auch später gezeigt hat, daß der Prozeß der Wirtszellreaktivierung sich nicht nur auf die Reparatur von UV-Schäden in der DNS beschränkt, so hat man sich doch bis zur jüngsten Zeit bei den Untersuchungen zur Aufklärung dieses Phänomens vorwiegend der strahlenschädigenden Wirkung des UV-Lichtes bedient. Für vergleichende Untersuchungen von Photoreaktivierung und Wirtszellreaktivierung war dies außerdem eine Notwendigkeit, da die PR wahrscheinlich ein spezifisch nur für die Reparatur von UV-Schäden entwickelter Mechanismus ist. Es ist daher verständlich, daß parallel zu diesen Untersuchungen der biologischen Strahlenwirkung ein intensives Studium der photochemischen Prozesse der Nucleinsäuren erfolgte. Die Ergebnisse dieser Studien sollen hier erwähnt werden, da sie zum Verständnis der folgenden Ausführungen notwendig sind[2].

[2] Ein Beitrag, der sich speziell mit der Wirkung des UV-Lichtes auf die Bakterienzelle befaßt, ist kürzlich erschienen (DOUDNEY, 1968).

1. Photoprodukte in Desoxyribonucleinsäuren

Als Folge der Einwirkung ultravioletter Strahlen auf DNS treten in der Hauptsache Dimerisationsprodukte von Pyrimidinbasen auf (Beukers und Berends, 1960; Wacker, 1963). In einem DNS-Molekül, das neben den Basen Adenin und Guanin die Pyrimidine Thymin und Cytosin enthält, entstehen Thymin-Thymin- (T=T, Abb. 1, I; Abb. 2), Thymin-Cytosin- (T=C, Abb. 1, II) und Cytosin-Cytosin-Dimere (C=C, Abb. 1, III). Die Cytosin enthaltenden Dimere desamiieren sehr leicht (in einer vom UV-Licht unabhängigen Reaktion) und gehen dabei in die entsprechenden Derivate des Uracils über (T=C→T=U; C=C→U=U; Abb. 3, I) (R. B. Setlow und Carrier, 1966). In einer DNS-Doppelhelix erfolgt die Dimerisation der in einer Helix benachbart stehenden Pyrimidinbasen wesentlich leichter als die zwischen Basen aus den benachbarten Helices. Im letzten Fall kommt es zu einer Vernetzung der beiden DNS-Stränge (Marmur und Grossman, 1961).

I.

II.

III.

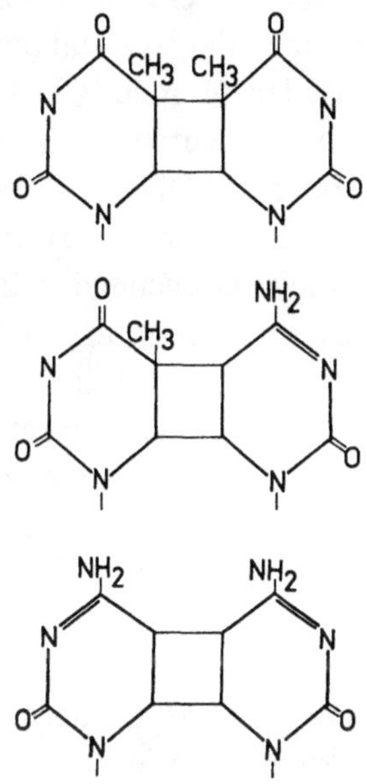

Abb. 1

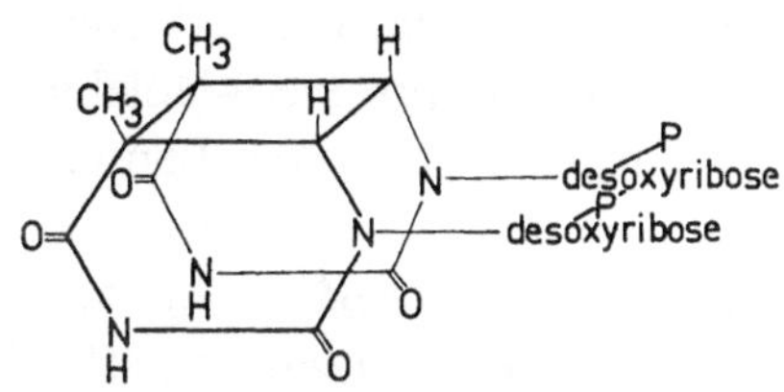

Abb. 2

Abb. 1. Photoprodukte in Desoxyribonucleinsäuren. Dimere Verbindungen von I. Thymin und Thymin (T=T), II. Thymin und Cytosin (T=C), III. Cytosin und Cytosin (C=C). (C-Atome im heterocyclischen Ring, sowie die an Ring-C- und Ring-N-Atome gebundenen H-Atome sind der Übersichtlichkeit wegen weggelassen)

Abb. 2. Das cis-syn-Dimere des Thymins in der Desoxyribonucleinsäure (modifiziert nach Weinblum, 1967.) (*P* Phosphorsäure)

Eine dimere Verbindung zwischen zwei Pyrimidinen kann in vier verschiedenen stereoisomeren Formen auftreten. Nach Bestrahlung von Thymidin bzw. dem Thymidylsäure-dinucleotid TpT sind tatsächlich auch alle Isomere gefunden worden. In der doppelsträngigen DNS tritt dagegen nur das in Abb. 2 wiedergegebene sog. cis-syn-Dimere auf (Weinblum und Johns, 1966; Weinblum, 1967). Seine Bildung ist nur möglich, indem die wendeltreppenartig übereinander angeordneten Basen in einem DNS-Strang ihre Lage zueinander verändern. Eine von zwei benachbarten Basen muß um 36° rotieren; außerdem muß der Abstand der Basen von 3,4 Å auf 1,54 Å verringert werden (Wang und Alcántara, 1965). Dies ist nur durch eine örtlich begrenzte Auflösung der Wasserstoffbrückenbindungen zwischen den komplementären

Basen der DNS-Doppelhelix möglich. Diese sekundäre Strukturveränderung mag im Zusammenhang mit dem Wirkungsmechanismus der an der Wirtszellreaktivierung beteiligten Enzyme von Wichtigkeit sein.

Die Bildung der Pyrimidin-Dimere ist eine photochemisch reversible Reaktion. Die Gleichgewichtslage ("photosteady state") zwischen Hin- und Rückreaktion ist in der Hauptsache von der Wellenlänge des einstrahlenden Lichtes abhängig (JOHNS et al., 1962). Die verschiedenen Basen und ihre entsprechenden Dimere haben bei einer gegebenen Wellenlänge verschiedene Absorptionskoeffizienten. Bei UV-Bestrahlung entstehen dadurch (pro entsprechendem Dinucleotid) unterschiedliche Mengen einzelner Dimertypen. Beispielsweise werden bei Bestrahlungen mit Licht der Wellenlängen 260 bis 280 mμ relativ mehr T=T als T=C und wiederum mehr T=C als C=C gebildet. Außerdem ist die relative Häufigkeit des Auftretens verschiedener Dimere abhängig von der Basenzusammensetzung der DNS. Die Wahrscheinlichkeit, mit der die Bildung eines bestimmten Dimeren zu erwarten ist, ist ungefähr proportional der Häufigkeit, mit der die entsprechenden Basen in der DNS *in Nachbarschaft* auftreten (R. B. SETLOW und CARRIER, 1966).

Ein Zusammenhang zwischen dem Auftreten von Pyrimidin-Dimeren in der DNS und der beobachteten biologischen Wirkung wird durch zahlreiche Befunde deutlich gemacht (R. B. SETLOW, 1967). 1. Es kommen auf eine letale Veränderung (Inaktivierung auf 1/e) eine genügende Zahl (ein oder mehrere) Dimere pro DNS-Molekül[3]. 2. Dimere inhibieren die DNS-Synthese *in vivo* und *in vitro*. Durch Photoreaktivierung werden die Dimeren wieder zu monomeren Basen revertiert, wobei die Synthesehemmung wieder aufgehoben wird. 3. Durch UV-Bestrahlung (280 mμ) entstandene Dimere lassen sich durch erneute Bestrahlung mit UV-Licht kürzerer Wellenlänge (240 mμ) wieder revertieren (Veränderung des "photosteady state"). Durch 280 mμ Bestrahlung inaktivierte Transformations-DNS läßt sich dadurch wieder reaktivieren ("short wavelength reactivation", s. III, 2). 4. Ersatz von Thymin durch Bromuracil verringert die Zahl der nach Bestrahlung gebildeten Dimere. Damit verbunden nimmt die Photo- bzw. Wirtszellreaktivierbarkeit der DNS ab.

Für UV-Bestrahlungen mit monochromatischem Licht werden Quecksilberhochdrucklampen verwendet (WINKLER et al., 1962). Die meisten Untersuchungen sind bisher aber mit Quecksilberniederdruckbrennern gemacht worden, die bevorzugt Licht der Wellenlänge 2537 Å emittieren (WACKER, 1963). Mit UV-Licht dieser Wellenlänge werden in der DNS von *E. coli*-Zellen Pyrimidin-Dimere mit einer Geschwindigkeit von $2{,}6 \times 10^{-6}$ Dimere/Thymin pro erg/mm² gebildet (RUPP und HOWARD-FLANDERS, 1968). Das sind 6,5 Dimere pro 10^7 Nucleotide (= ungefähre Gesamtzahl an Nucleotiden pro Bakteriengenom). Diese Daten beruhen auf einer Abschätzung der relativen

[3] Dies gilt strenggenommen nur für biologische Systeme, in denen doppelsträngige DNS vorkommt. Beim Bakteriophagen Φ X 174, der einsträngige DNS enthält, sind nur 0,34 T=T pro letalem „Treffer" gemessen worden (DAVID, 1964). Hier müssen andere Photoprodukte als T=T-Dimere für die Inaktivierung verantwortlich gemacht werden.

Häufigkeit, mit der T=T, T=C und C=C in *E. coli*-DNS bei kleinen UV-Dosen (2×10^3 erg/mm^2) gebildet werden, nämlich zu 50%, 40% bzw. 10% (R. B. SETLOW und CARRIER, 1966).

Die quantitative Bestimmung der Dimere wurde bisher so durchgeführt, daß das bestrahlte Objekt (Bakterien oder Viren oder die isolierte DNS aus denselben) einer Säurehydrolyse bei Temperaturen von 100° C oder darüber unterzogen wurde. Die dimeren Verbindungen sind säurestabil und können später papierchromatographisch analysiert werden (WACKER, 1963). Neuere Untersuchungen haben aber Zweifel aufkommen lassen, ob die nach Säurehydrolyse gefundenen Dimeren wirklich die primären Reaktionsprodukte in der DNS darstellen (PEARSON et al., 1965; WANG und VARGHESE, 1967). Außerdem sind bei den bisherigen Analysen weitere UV-Produkte dadurch übersehen worden, daß diese bei den bisherigen Verfahren nicht von den monomeren bzw. dimeren Basen getrennt wurden (YAMANE et al., 1967). Es muß deshalb damit gerechnet werden, daß einige bisher veröffentlichte Daten, die Korrelation von biologischer Wirkung und photochemischer Veränderung betreffend, einer Überprüfung unterzogen werden müssen.

2. Photoprodukte in Ribonucleinsäuren

Die photochemischen Prozesse, die sich bei UV-Bestrahlung von biologisch aktiven Ribonucleinsäuren abspielen, sind bisher nicht so eingehend untersucht worden, wie die bei der DNS. Dieser Mangel ist hauptsächlich darauf zurückzuführen, daß in der RNS qualitativ mehr, und vor allem instabilere Photoprodukte gebildet werden, deren Nachweis oftmals schwierig ist. Während die Pyrimidinbasen in beiden Nucleinsäuretypen (DNS und RNS) dimerisieren können, besteht hinsichtlich der durch UV-Licht induzierten Bildung von Pyrimidinhydraten (Abb. 3, II, III) ein grundlegender Unterschied zwischen DNS- und RNS-Molekülen. Die in der RNS oder in Homopolymeren von Ribonucleotiden vorkommende Base Uracil kann — neben der Dimerisierung mit einer benachbarten Pyrimidinbase — auch ein Hydrat bilden (Abb. 3, II) (McLAREN und SHUGAR, 1964). Diese Reaktion ist bei der analogen Base Thymin in der DNS nicht möglich. Ebenso wie das Uracil kann auch die Base Cytosin dimerisieren oder ein Hydrat bilden (Abb. 3, III). Das Cytosinhydrat ist aber im Vergleich zum Uracilhydrat sehr instabil. Durch Dehydratation wird es zum größten Teil wieder in Cytosin zurückverwandelt, zum geringeren Teil wird es zu Uracilhydrat desaminiert (SCHUSTER, 1964; HARIHARAN und JOHNS, 1968). Die Hydratbildungen des Uracils und des Cytosins sind — im Gegensatz zur Dimerisierung — *photo*chemisch irreversible Reaktionen. Bei UV-Bestrahlung von (einsträngiger) Polycytidylsäure (GRIFFITH und R. B. SETLOW, 1965), nicht aber bei den Doppelstrang-Homopolymeren aus Desoxycytidylsäure und Desoxyinosylsäure (R. B. SETLOW et al., 1965) [bzw. dem entsprechenden Homopolymeren aus Cytidylsäure und Inosylsäure (WIERZCHOWSKI und SHUGAR, 1962)] ist die Bildung von Cytosinhydrat, wenn auch indirekt, nachgewiesen worden. Die räumliche Fixierung der Cytosinbasen in der Doppelhelix verhindert oder beeinträchtigt

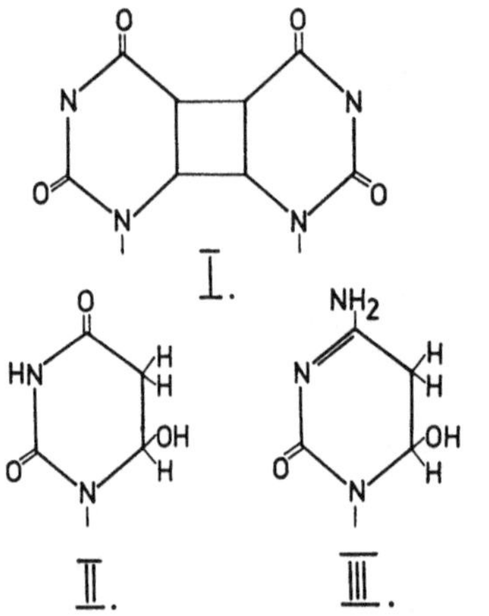

Abb. 3. Photoprodukte in Ribonucleinsäuren. I. Dimere Verbindung des Uracils (U=U). (C- und H-Atome weggelassen wie in Abb. 1 angegeben). II. Uracilhydrat, III. Cytosinhydrat

zumindest deren Hydratbildung. Vermutlich gilt diese unterschiedliche Reaktionsfähigkeit auch für die Cytosinbasen in den natürlich vorkommenden Nucleinsäuren. Man muß deshalb beachten, daß bei einsträngigen DNS- und RNS-Molekülen, kaum aber bei Doppelstrang-DNS, eine beobachtete biologische Wirkung unter Umständen auch auf die Bildung eines Cytosinhydrats (bzw. dessen Desaminierungsprodukt) zurückzuführen sein kann.

Das Auftreten von dimerisierten *und* hydratisierten Pyrimidinbasen in der RNS macht die Zuordnung der biologischen Wirkung zu einer photochemischen Veränderung wesentlich schwieriger als in der DNS. So konnte bis heute noch nicht eindeutig geklärt werden, welche photochemischen Reaktionen für die Inaktivierung bzw. Photoreaktivierung von biologisch aktiven Ribonucleinsäuren verantwortlich sind (SARIN und JOHNS, 1968; MERRIAM und GORDON, 1967).

III. Nicht-enzymatische Photoreaktivierung

1. Definition

Unter Photoreaktivierung (PR) versteht man eine lichtabhängige Regeneration von strahlengeschädigten biologischen Systemen. Der Schaden (die Inaktivierung) wird durch Bestrahlung mit UV-Licht (<3000 Å) verursacht. Seine Eliminierung (Reaktivierung) erfolgt durch *gleichzeitige* oder *nachfolgende* Bestrahlung mit Licht *anderer* als der zur Inaktivierung verwendeten Wellenlänge. Mit dieser Definition (JAGGER und STAFFORD, 1965a) läßt sich die PR unterscheiden von dem Phänomen der sog. Photoprotektion (WEATHERWAX, 1956; JAGGER, 1960), bei dem die Wirkung inaktivierender Strahlen durch *vorausgehende* Bestrahlung mit Licht längerer Wellenlänge abgeschwächt wird. Bei der PR selbst unterscheidet man zwischen den *direkten* und den *indirekten* Reaktivierungsprozessen.

Bei der direkten PR wird der Strahlenschaden sofort und direkt durch die Energie des photoreaktivierenden Lichtquants eliminiert. Dies kann geschehen durch UV-Licht von *kürzerer* als der zur Inaktivierung verwendeten Wellenlänge ["short wavelength reactivation" (R. B. SETLOW und J. K. SETLOW, 1962), meistens UV-Licht der Wellenlänge um 2400 Å] oder durch Licht längerer Wellenlänge (3100—5000 Å) unter gleichzeitiger Mitwirkung eines photoreaktivierenden Enzyms (RUPERT et al., 1958; RUPERT, 1960a). Die indirekte PR hingegen ist ein zeitlich und/oder örtlich von der Wirkung des

Lichts entfernter Prozeß, bei dem ein photoreaktivierendes Enzym nicht beteiligt ist (JAGGER und STAFFORD, 1965a).

Wenngleich in diesem Beitrag nur das Phänomen einer direkten, unter Beteiligung eines Enzyms ablaufenden PR näher besprochen werden soll, so erscheint es doch notwendig, die verschiedenen Reaktivierungsprozesse einleitend kurz zu charakterisieren. Bei der PR einer strahlengeschädigten Zelle ist es nämlich oftmals schwierig, wenn nicht unmöglich, herauszufinden, in welchem Maße direkte und indirekte Reaktivierungsprozesse zu einem beobachteten Effekt beigetragen haben (RUPERT und HARM, 1966).

2. "Short wavelength reactivation"

Transformations-DNS wird durch UV-Licht inaktiviert. Die Inaktivierungskinetik ist abhängig von der Wellenlänge des UV-Lichtes. Auf dieselbe UV-Dosis bezogen, inaktiviert Licht der Wellenlänge 2850 Å die DNS schneller als noch kurzwelligeres Licht (2400 Å) (J. K. SETLOW, 1966b). Eine durch 2850 Å Bestrahlung inaktivierte DNS kann durch Licht der Wellenlänge 2400 Å wieder reaktiviert werden (Abb. 4) (R. B. SETLOW und SETLOW, 1962; J. K. SETLOW, 1966b). Inaktivierung und Reaktivierung werden beide nach Bestrahlung *in vitro* erzielt. Der Effekt beruht lediglich auf einer Veränderung des Gleichgewichtes ("photosteady state") zwischen sich bildenden und zerfallenden Pyrimidin-Dimeren in der DNS. Im Gleichgewicht ist die Zahl der Dimere (pro monomere Base) bei 2850 Å größer als bei 2400 Å. Eine Reaktivierung kann nur dann beobachtet werden, wenn die Zahl der bei der Inaktivierung (2850 Å) gebildeten Dimere erheblich größer ist als die im "photosteady state" *bei 2400 Å-Bestrahlung* vorhandene Zahl an Dimeren. Sie ist in jedem Falle gering und führt bei fortgesetzter Bestrahlung mit kurzwelligem UV-Licht zu einer erneuten Inaktivierung (Abb. 4).

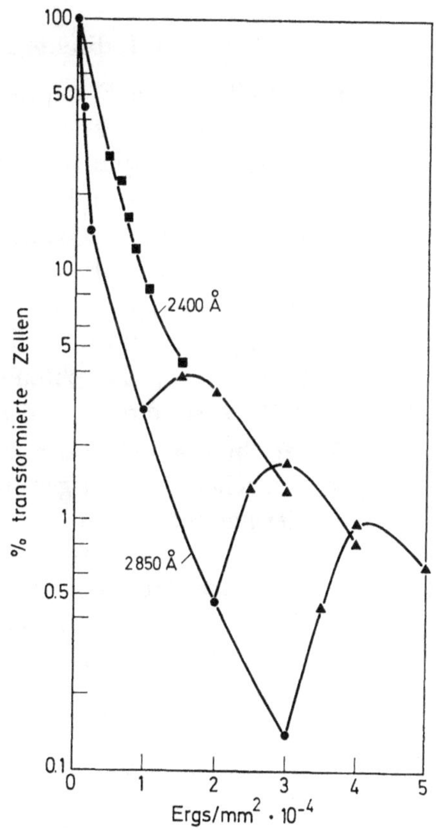

Abb. 4. "Short wavelength reactivation" von Transformations-DNS. Transformations-DNS (Bacillus subtilis) wird mit monochromatischem UV-Licht bestrahlt. -●- 2850 Å-Bestrahlung allein, -■- 2400 Å-Bestrahlung allein, -▲- 2850 Å-Bestrahlung, gefolgt von 2400 Å-Bestrahlung; die angegebene Dosis stellt die Summe der 2850 Å- und der 2400 Å-Dosen dar. (BOLING und SETLOW, unveröffentlicht; nach J. K. SETLOW, 1966b)

Die "short wavelength reactivation" wird noch etwas effektiver, wenn der inaktivierten DNS vor der 2400 Å Bestrahlung Proflavin zugesetzt wird (J. K. SETLOW und SETLOW, 1967). Der Acridinfarbstoff lagert sich offenbar zwischen zwei Nucleotidpaare mit benachbarten Thyminbasen und verhindert deren Dimerisierung durch UV-Bestrahlung (LERMAN, 1961; BEUKERS, 1965), ohne die Spaltung der bei 2850 Å schon gebildeten Dimere zu beeinflussen (R. B. SETLOW und CARRIER, 1967).

Der Prozeß der "short wavelength reactivation", der bisher nur bei Transformations-DNS, nicht aber bei UV-Bestrahlung von Bakterien beobachtet worden ist (J. K. SETLOW, 1966b), macht deutlich, daß Dimerisationsprodukte in der bestrahlten DNS zur Inaktivierung beitragen (s. II, 1). Es ist geschätzt worden, daß zwischen 50 und 70% der durch langwelliges UV-Licht induzierten Letalschäden auf die Bildung dimerer Basen zurückzuführen sind (J. K. SETLOW und SETLOW, 1963).

3. Indirekte Photoreaktivierung

Nach Behandlung von *E. coli*-Zellen mit einem mutagenen Agenz (HNO_2) isolierten HARM und HILLEBRANDT (1962) eine Mutante, die kein aktives photoreaktivierendes Enzym mehr bildet. Jedoch konnten nach UV-Inaktivierung nicht nur Wildtyp- (phr^+) sondern auch Mutantenzellen (phr^-) durch Belichtung (Wellenlänge >3000 Å) wieder reaktiviert werden (JAGGER und STAFFORD, 1965a). Ein Reaktivierungsprozeß, der vom photoreaktivierenden Enzym unabhängig ist, war damit nachgewiesen, und man bezeichnete ihn als indirekte PR. Während die PR der phr^--Mutante ausschließlich auf der Wirkung dieses einen, indirekten Prozesses beruht, können im Wildtyp (phr^+) beide Prozesse, direkte und indirekte PR, zur Wirkung kommen.

Der Unterschied zwischen der direkten enzymatischen und der indirekten PR ist in Versuchen deutlich geworden, bei denen monochromatisches Licht zur Reaktivierung benutzt wurde (JAGGER und STAFFORD, 1965a). Licht der Wellenlänge 3341 Å reaktiviert sowohl UV-inaktivierte Zellen des Wildtyps wie auch die der phr^--Mutante, wenn auch mit unterschiedlicher Wirksamkeit; der Effekt ist pro UV-Dosis größer im Wildtyp als in der Mutante. Jedoch erfolgt bei 4047 Å Belichtung nur eine PR des Wildtyps, nicht aber bei der Mutante. Andererseits zeigen die Aktionsspektren für die *in vitro* Reaktivierung von UV-inaktivierter Transformations-DNS mit photoreaktivierendem Enzym (Extrakte von *E. coli*- bzw. Hefezellen) Maxima bei etwa 3900 Å und eine geringere Höhe unterhalb 3500 Å (J. K. SETLOW und BOLING, 1963; J. K. SETLOW, 1966b). Man kann daraus schließen, daß in der strahlengeschädigten Wildtypzelle die PR bei Licht der Wellenlänge um 4000 Å direkt, d. h. allein durch die Wirkung des photoreaktivierenden Enzyms zustande kommt, während sich bei Wellenlängen um 3400 Å direkte und indirekte PR überlappen.

Die *in vitro* PR von Transformations DNS mit photoreaktivierendem Enzym (Wellenlänge >3100 Å) ist von der Temperatur und der pro Zeiteinheit eingestrahlten Lichtdosis abhängig. Mit steigender Temperatur wird auch die Geschwindigkeit der PR größer. Bei einer gegebenen, suboptimalen Lichtdosis ist die PR größer, wenn pro Zeiteinheit weniger Lichtquanten eingestrahlt werden (RUPERT, 1964a). Bestimmt man diese Parameter bei der PR von *E. coli*-Zellen, so ergibt sich folgendes Bild (JAGGER und STAFFORD, 1965a). Beim Wildtyp (phr$^+$) ist bei 4047 Å-Belichtung die Reaktivierung temperatur- und Lichtdosis-abhängig, in Übereinstimmung mit der *in vitro*-Reaktion. Hingegen wird bei 3341 Å-Belichtung bei der Mutante (phr$^-$) keine, beim Wildtyp nur eine geringe Abhängigkeit von der Lichtdosis beobachtet. Ähnliches gilt für die Temperaturabhängigkeit der 3341 Å-Photoreaktivierung beider Bakterienstämme. Die indirekte PR ist somit unabhängig von der Lichtdosis und der Temperatur; außerdem kommt bei 3341 Å-Belichtung des Wildtyps die indirekte PR gegenüber der enzymatischen PR stärker zum Ausdruck. Wie in Abschnitt IV, 2 noch gezeigt werden wird, werden bei dem Prozeß der enzymatischen PR die in der DNS gebildeten dimeren Basen wieder monomerisiert (COOK, 1967). Hingegen wird bei der indirekten PR keine Spaltung der Dimeren beobachtet (JAGGER und STAFFORD, 1965b).

Die indirekte PR ist dem im Abschnitt III, 1 erwähnten Phänomen der Photoprotektion sehr ähnlich. Dies gilt insbesondere für die Wellenlängenabhängigkeit beider Prozesse (JAGGER et al., 1964). So zeigt das Aktionsspektrum der Photoprotektion ein Maximum bei 3341 Å, d.h. eine der inaktivierenden UV-Strahlung (meist 2537 Å) vorausgehende Belichtung von Bakterienzellen schwächt den UV-strahlenschädigenden Effekt am stärksten ab, wenn Licht der Wellenlänge 3341 Å eingestrahlt wird. Hingegen hat die vorausgehende Belichtung bei 4047 Å keinen abschwächenden Effekt mehr auf die nachfolgende UV-Bestrahlung.

Da die Belichtung von normalen, nicht UV-inaktivierten *E. coli*-Zellen eine Wachstums- und Teilungsverzögerung hervorruft, deren Aktionsspektrum ebenfalls ein Maximum bei 3341 Å aufweist, wird vermutet, daß dieser Prozeß mit der Photoprotektion und mit der indirekten PR ursächlich zusammenhängt (JAGGER et al., 1964). Man vermutet, daß durch die lichtinduzierte Wachstumsverzögerung andere, lichtunabhängige Reaktivierungsprozesse in der Zelle (z.B. die Wirtszellreaktivierung) effektiver arbeiten können, weil diesen jetzt — im Vergleich zu der Situation in normal wachsenden Zellen — mehr Zeit zur Elimination von Strahlenschäden zur Verfügung steht (JAGGER und STAFFORD, 1965a).

Tatsächlich beobachtet man auch bei *E. coli*-Zellen, deren DNS-Replikation künstlich verzögert wurde, eine wesentlich geringere Empfindlichkeit gegenüber UV-Strahlen als bei Zellen in der exponentiellen Wachstumsphase (HANAWALT, 1966). Jedoch ist bei der 3341 Å-Belichtung von *E. coli* zwar eine Hemmung der RNS- und Proteinsynthese, nicht aber der DNS-Synthese beobachtet worden (J. K. SETLOW, 1966a). Eine Aufklärung der vermuteten Zusammenhänge wird möglich sein, wenn Bakterienmutanten zur Verfügung stehen, die sowohl in der Photo- wie auch in der Wirtszellreaktivierung defizient sind.

Die Entdeckung der indirekten PR ist relativ jüngeren Datums (JAGGER und STAFFORD, 1965a). Daher bedürfen sicher viele der früher bekanntgewordenen Daten über die in Zellen beobachtete PR einer kritischen Überprüfung. Die Aussage, daß eine direkte, enzymatische PR einer beobachteten Reaktivierung zugrundeliegt, kann mit Sicherheit nur dann gemacht werden, wenn es gelingt, die PR auch *in vitro* nachzuweisen. Es ist sicher auszuschließen, daß eine indirekte PR in einem zellfreien System stattfinden kann.

IV. Enzymatische Photoreaktivierung

1. Vorkommen und Nachweis

Eine lichtabhängige *in vitro*-Reaktivierung von UV-inaktivierter Transformations-DNS (aus Haemophilus influenzae) wurde erstmalig mit einem Extrakt aus *E. coli* erzielt (RUPERT et al., 1958). Später wurde eine aktive Komponente auch aus Hefezellen extrahiert (RUPERT, 1960a). In jüngster Zeit ist ein photoreaktivierendes Enzym auch in Extrakten von Blaualgen (WERBIN und RUPERT, 1968) und Neurospora crassa (TERRY und SETLOW, 1967), sowie in Gewebehomogenaten von Echinodermen, Arthropoden und einigen Vertebraten (COOK und MCGRATH, 1967; REGAN und COOK, 1967) gefunden worden. Allerdings ist in Säugetierzellen noch keine photoreaktivierende Aktivität beobachtet worden (CLEAVER, 1966). Auch in Extrakten dieser Zellen konnte bisher kein photoreaktivierendes Enzym nachgewiesen werden (COOK und MCGRATH, 1967).

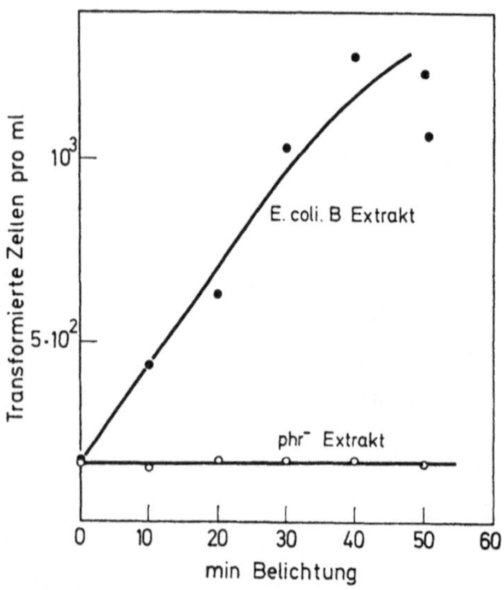

Abb. 5. In vitro-Photoreaktivierung von UV-inaktivierter Transformations-DNS mit einem Extrakt aus *E. coli* B. UV-bestrahlte Transformations-DNS (Haemophilus influenzae) wird unter Belichtung mit Extrakten von *E. coli* B phr^+ und *E. coli* phr^- (HARM und HILLEBRANDT, 1963) behandelt. (Nach RUPERT, 1965)

In allen Fällen besteht der Nachweis eines photoreaktivierenden Enzyms darin, daß den Zellextrakten zugesetzte UV-inaktivierte Transformations-DNS (RUPERT, 1962a, b) oder Virus-DNS (HARM und RUPERT, 1963) in Gegenwart von Licht (>3000 Å) wieder reaktiviert wird. Dabei können bis zu 90% der UV-Schäden eliminiert werden. Parallel zu der Reaktivierung muß die Zahl der Pyrimidin-Dimere abnehmen. Die aktive Komponente muß durch Hitze oder Verdauung mit proteolytischen Enzymen inaktivierbar sein. Bei Zellextrakten, in denen keine enzymatische Aktivität nachzuweisen ist, setzt man photoreaktivierendes Enzym aus Hefe zu und wiederholt den Test, um auszuschließen, daß der Extrakt Hemmstoffe enthält und dadurch eine PR maskiert. Abb. 5 zeigt als Beispiel die *in vitro*-PR von Transformations-DNS aus H. influenzae durch einen Extrakt aus *E. coli*. Die phr$^-$-Mutante desselben Stammes liefert einen inaktiven Extrakt (RUPERT, 1965).

Die photoreaktivierenden Enzyme aus *E. coli* und Hefe überlappen sich vollständig in ihrer Wirkung, d.h. Extrakte aus beiden Organismen photoreaktivieren UV-inaktivierte DNS gleich gut, bzw. durch das *E. coli*-Enzym maximal photoreaktivierte DNS wird durch das Enzym aus Hefe nicht weiter reaktiviert und umgekehrt (RUPERT, 1964a). Ursprünglich hatte man vermutet, daß sich das PR-System aus *E. coli* im Gegensatz zu dem aus Hefe aus zwei Komponenten, einer nicht dialysierbaren, hitzelabilen Fraktion und einem dialysierbaren, hitzestabilen Faktor zusammensetze. Wahrscheinlicher ist aber, daß Rohextrakte aus Bakterien dialysierbare Nuclease-Inhibitoren enthalten. Bei deren Abwesenheit würde Transformations-DNS allmählich durch Nucleasen abgebaut und dadurch eine geringere PR vortäuschen (RUPERT, 1964a). Damit der Nachweis eines photoreaktivierenden Enzyms nicht durch die mögliche Gegenwart von Nucleasen beeinträchtigt wird, setzt man neuerdings der zu reaktivierenden Transformations-DNS unbestrahlte DNS aus anderer Quelle zu. Letztere stört die Reaktion zwischen UV-bestrahlter DNS und photoreaktivierendem Enzym nicht, schirmt die bestrahlte DNS aber gegen Nucleasewirkung ab, da sie selbst als Substrat dient (COOK und MCGRATH, 1967).

Die Schwierigkeit des Arbeitens mit Extrakten aus Bakterien hat dazu geführt, daß die meisten Untersuchungen mit dem Enzym aus Hefe durchgeführt wurden. Andererseits sind aber *in vivo*-Studien der PR vorwiegend mit Bakterien gemacht worden. Es fehlen deshalb noch vergleichende Untersuchungen zwischen den *in vivo* und *in vitro* gemessenen Photoreaktivierungsprozessen, Studien, die zur Aufklärung der Funktion und Wirkungsweise des photoreaktivierenden Enzyms unerläßlich sind.

2. Mechanismus der Enzymwirkung

Photoreaktivierendes Enzym aus Hefe bindet sich an UV-bestrahlte, nicht aber an unbestrahlte Transformations-DNS. Diese Bindung ist im Dunkeln stabil, bei Belichtung dissoziieren DNS und Enzym wieder. Das photoreaktivierende Enzym sedimentiert bei der Ultrazentrifugation wesentlich langsamer als DNS, hingegen bewegt sich der Komplex aus bestrahlter DNS und dem Enzym als eine Komponente und mit etwa der Geschwindigkeit von unbestrahlter DNS. Er läßt sich ferner im Dunkeln undissoziiert über Sephadex-Gel filtrieren (RUPERT, 1962b).

Der gebundene Zustand verleiht dem Enzym besondere Eigenschaften. Seine Empfindlichkeit gegenüber Inaktivierung durch Erhitzen auf 65° wird drastisch reduziert. Seine Komplexbildung mit der DNS stabilisiert es außerdem gegen den Angriff von Schwermetallionen. Erwerb und Verlust dieser Eigenschaften sind reversibel wie die Bindung des Enzyms an die DNS und seine Dissoziation durch Belichtung (RUPERT, 1962b). Da nur während der Belichtung des Komplexes die biologische Aktivität der Transformations-DNS regeneriert wird, muß die Dissoziation von der Reaktivierung begleitet sein. Diese Daten erlauben es, die Photoreaktivierung als einen enzymatischen Prozeß zu behandeln (RUPERT, 1962a), der dem Michaelis-Menten-Schema folgt (FRUTON und SIMMONDS, 1958):

$$E + S \underset{k_2}{\overset{k_1}{\rightleftharpoons}} ES \xrightarrow{k_3 (h\nu)} E + P$$

In dieser Gleichung bedeuten E photoreaktivierendes Enzym, S „Substrat" (UV-Schaden in der DNS), ES der Komplex, P eine regenerierte DNS-Struktur. k_1, k_2 und k_3 sind Geschwindigkeitskonstanten, wobei k_3 eine lichtabhängige Konstante ist (mit $k_3 = 0$ im Dunkeln).

Als „Substrat" zählen die UV-Schäden in den DNS-Molekülen. Sie sind nicht einheitlich in ihrer chemischen Struktur und ihre Zahl ist in dem Inaktivierungsbereich, in dem eine PR gemessen wird, gering. Außerdem sind in einer Population von Nucleinsäuremolekülen die Schäden statistisch verteilt. Die „Substrat"-Konzentration wird deshalb kaum mit Genauigkeit zu bestimmen sein (RUPERT, 1964a). Jedoch entspricht die Kinetik der PR, wenn die Konzentrationen an Enzym und UV-bestrahlter DNS variiert werden, in der erwarteten Weise dem obigen Reaktionsschema.

Der Enzym-Substrat-Komplex muß die zu seiner Dissoziation erforderlichen Lichtquanten absorbieren. Wie im Abschnitt III, 3 erwähnt wurde, zeigt das Aktionsspektrum für die *in vitro*-PR ein Maximum bei etwa 3900 Å. Gereinigtes photoreaktivierendes Enzym aus Hefe (etwa 3000fache Anreicherung) besitzt neben dem für Proteine typischen Absorptionsmaximum bei 2800 Å eine geringe, für Cytochrome typische Absorption bei 4150 Å (MUHAMMED, 1966). Die Verbindung, die diese Absorption besitzt, läßt sich aber im Verlauf der Enzymreinigung weitgehend abtrennen. Es ist deshalb noch offen, ob es sich bei dem beim Enzym verbleibenden Material tatsächlich um den Lichtacceptor oder lediglich um eine Verunreinigung handelt.

Die Möglichkeit besteht auch, daß sich die lichtabsorbierende Komponente erst durch die Vereinigung von Enzym und Substrat ausbildet. In diesem Falle dürfte ihr Nachweis schwierig sein. Die Tatsache, daß sich das Spektrum einer *Mischung* von UV-bestrahlter DNS (aus Kalbsthymus) und photoreaktivierendem Enzym (aus Hefe) oberhalb von 3000 Å nicht von dem des gereinigten Enzyms unterscheidet (MUHAMMED, 1966), spricht noch nicht gegen diese Möglichkeit; bei der angegebenen Spektrumsmessung war nämlich die Menge des an die DNS *gebundenen* PRE nicht bestimmt worden.

Das Produkt P der enzymatischen Reaktion ist die regenerierte DNS-Struktur. Da die zuvor inaktivierte DNS ihre biologische Aktivität durch PR wiedergewinnt, ist anzunehmen, daß dabei auch ihre ursprüngliche Struktur wiederhergestellt wird. Die durch UV-Bestrahlung induzierten Pyrimidin-Dimere müssen also wieder gespalten werden. Zunächst hatte man nach PR nur das Verschwinden von Thymin-Dimeren in der DNS festgestellt (WULFF und RUPERT, 1962; R. B. SETLOW et al., 1963). Lediglich bei einem enzymatisch synthetisierten Polymeren, dem poly dI:dC (mit radioaktiv markiertem

Cytosin), war nach UV-Bestrahlung und PR eine Spaltung von Dimeren nachgewiesen worden. Die in diesem Polymeren UV-induzierten Cytosin-Dimere lassen sich durch Erwärmen leicht in Uracil-Dimere umwandeln. Bei der anschließenden PR war deren Verschwinden begleitet von der Entstehung von radioaktivem Uracil, das sich neben dem noch verbliebenen Cytosin leicht nachweisen ließ (R. B. SETLOW et al., 1965). In jüngster Zeit ist auch der Nachweis gelungen, daß die in natürlicher DNS gebildeten Thymin-Dimere bei der PR wieder in zwei Thyminbasen gespalten werden (COOK, 1967). Wird UV-bestrahlter ^{3}H-Thymin-markierter DNS ^{14}C-Thymin-markierte DNS zugesetzt, so stellt man nach PR der Mischung beider Nucleinsäuren eine Erhöhung des ^{3}H/^{14}C-Thymin-Verhältnisses fest. Dabei ist die bei der PR beobachtete Abnahme der Thymin-Dimeren begleitet von einer stöchiometrischen Zunahme des Thymins.

3. Substrat und kompetitive Hemmung

Alle natürlich vorkommenden oder künstlich erzeugten DNS-Moleküle, die bei UV-Bestrahlung Pyrimidin-Dimere bilden können, sollten als Substrat für das photoreaktivierende Enzym in Frage kommen. Dies ist mit gewissen Einschränkungen auch der Fall (RUPERT, 1964b).

Biologisch aktive Desoxyribonucleinsäuren geben durch ihre Photoreaktivierbarkeit ihre Substratnatur zu erkennen. Inwieweit jedoch biologisch inaktive Moleküle oder enzymatisch synthetisierte DNS-Polymere als Substrate in Frage kommen, war zunächst schwierig herauszufinden. Die kompetitive Hemmbarkeit des photoreaktivierenden Enzyms aus Hefe, nach dem Michaelis-Menten-Reaktionsschema (Abschnitt IV, 2) zu erwarten, hat sich später als Nachweismethode für den Substratcharakter einer UV-bestrahlten Verbindung benutzen lassen (RUPERT, 1961).

Die Reaktionsgeschwindigkeit der PR von UV-bestrahlter Transformations-DNS ist dieselbe, ob diese allein oder zusammen mit unbestrahlter, nicht-transformierender DNS mit dem PRE inkubiert wird. Werden hingegen äquivalente Mengen von Transformations DNS und nicht-transformierender DNS, die beide dieselbe UV-Dosis erhalten haben, gemeinsam photoreaktiviert, so sinkt die Reaktionsgeschwindigkeit auf die Hälfte. Dagegen wird keine Reduktion der Geschwindigkeit der PR beobachtet, wenn die bestrahlte, nicht-transformierende DNS erst allein mit dem Enzym im Licht inkubiert und dann der inaktivierten Transformations-DNS zugesetzt wird (RUPERT, 1962a). Diese Versuche zeigen, daß Transformations-DNS und biologisch inaktive DNS in eine Konkurrenzreaktion um die vorhandenen Enzymmoleküle treten. An Stelle der biologisch inaktiven DNS kann jede beliebige DNS verwendet werden, bei der photoreaktivierbare UV-Schäden nachgewiesen werden sollen.

Als zusätzliches Indiz für das Vorhandensein einer kompetitiven Hemmung kann die größere Hitzeresistenz des photoreaktivierenden Enzyms im Enzym-Substrat-Komplex angesehen werden. Wenn Enzymmoleküle einmal mit

Tabelle 1. *Kompetitive Inhibitoren für das photoreaktivierende Enzym aus Hefe*

UV-bestrahltes Substrat	Kompetitive Hemmung[1]
DNS[2] aus H. influenzae, *E. coli*, T_2, Kalbsthymus, ΦX 174[a]	+
Native oder denatuierte DNS des Phagen SP2 [3,f]	+
Hitzedenaturierte DNS [4,g]	+
Poly dA:dT[b]	+
Poly dAT:dAT [5,c]	—
Poly dG:dC [d], poly dI:dC [e], poly dC [b]	+
Oktathymidylsäure $(pT)_8$, TpT[6], Thymidin[6], Thymin[6,d]	—
Apurinsäure	—
RNS, poly A, poly C, poly U, poly A:U[a]	—
BU enthaltende DNS[h]	+

[1] + (—) bedeutet, daß kompetitive Hemmung (nicht) beobachtet wurde.
[2] Als Beispiele für natürlich vorkommende Desoxyribonucleinsäuren. Solche mit hohem AT-Gehalt sind (pro UV-Dosis) bessere Inhibitoren als Moleküle mit hohem GC-Gehalt (J. K. SETLOW, 1966a).
[3] Der Phage SP2 enthält Uracil an Stelle von Thymin.
[4] Die UV-Bestrahlung erfolgte vor der Hitzedenaturierung (J. K. SETLOW, 1966a).
[5] Nach UV-Bestrahlung konnten spektrophotometrisch keine Thymin-Dimere nachgewiesen werden (DEERING und SETLOW, 1963).
[6] Die dimere Verbindung des Moleküls wurde im Test verwendet.

[a] RUPERT (1964a).
[b] J. K. SETLOW et al. (1965).
[c] RUPERT (1961).
[d] RUPERT (1964b).
[e] R. B. SETLOW et al. (1965).
[f] RUPERT und HARM (1966).
[g] J. K. SETLOW (1966a).
[h] RUPERT (1961; 1966).

einem kompetitiven Inhibitor zusammen, das andere Mal allein (bei Dunkelheit) erhitzt und anschließend zur PR von Transformations-DNS verwendet werden, so ist die *verbliebene* Enzymaktivität im ersten Falle höher und damit die Geschwindigkeit der PR der Transformations-DNS auch größer (RUPERT, 1962b).

Tabelle 1 gibt eine Übersicht über die Eigenschaften UV-bestrahlter Nucleinsäuren und deren Derivate als kompetitive Hemmstoffe (und damit auch als Substrate) für das photoreaktivierende Enzym aus Hefe. Demnach kommen Doppelstrang- wie auch Einzelstrang-DNS, nicht aber Ribonucleinsäure oder synthetische Polyribonucleotide als Enzyminhibitoren in Frage (RUPERT, 1964b). Die beiden DNS-Typen haben eine deutlich verschiedene Hemmwirkung. Bestrahlte DNS, durch Erhitzen auf 100°C und rasches Abkühlen denaturiert, hemmt nur etwa halb so gut wie dieselbe DNS nicht denaturiert. In Übereinstimmung damit findet man einmal eine langsamere und außerdem unvollständige Eliminierung der Thymin-Dimeren in denaturierter DNS durch das photoreaktivierende Enzym (Abb. 6), (R. B. SETLOW, 1964), zum anderen sind bei den DNS-Einzelstrang-Phagen ΦX 174 und S 13 nur zwischen 20 und 30 % der UV-Schäden photoreaktivierbar [verglichen mit bis zu 90 % bei Doppelstrang-DNS (HARM, 1967)]. Alle enzymatisch synthetisierten DNS-Polymere, sofern sie Pyrimidinbasen in Nachbarschaft enthalten, wirken auch als Inhibitoren. Voraussetzung hierfür ist aber, daß sie eine Mindestgröße besitzen. Bei UV-bestrahlten Oligodeoxythymidylsäuren nimmt die kompetitive Hemmung mit kleiner werdender Kettenlänge ab. Das bestrahlte Oktanucleotid $(pT)_8$ tritt schon nicht mehr in Wechselwirkung mit dem photoreaktivierenden Enzym (J. K. SETLOW u. F. J. BOLLUM, 1968; RUPERT, 1964b). Ebenfalls unwirksam ist die Apurinsäure, obwohl hier nach UV-Bestrahlung und Hydrolyse Thymin-Dimere nachgewiesen wurden. Offenbar muß eine Polynucleotidkette intakt sein, damit sich das Enzym anlagern kann (RUPERT, 1964b).

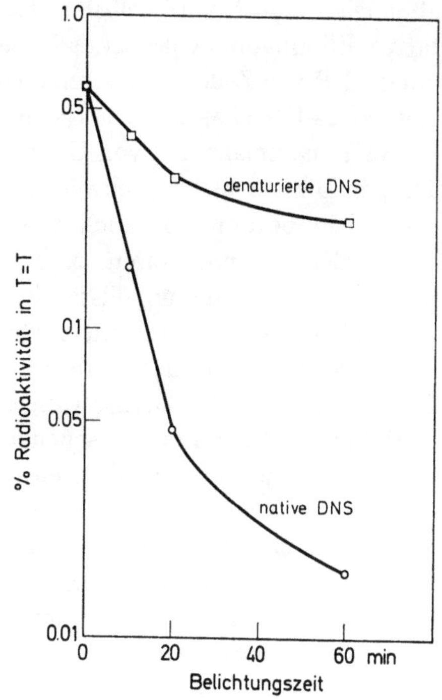

Abb. 6. Spaltung von Thymin-Dimeren in E. coli DNS durch das photoreaktivierende Enzym aus Hefe. Native und hitzedenaturierte *E. coli*-DNS (^{3}H-markiert) wird mit photoreaktivierendem Enzym aus Hefe unter Belichtung (3300—4400 Å) inkubiert. Gemessen wird die in der DNS verbleibende Menge an (radioaktiven) Thymidin-Dimeren. (R. B. SETLOW und CARRIER, unveröffentlicht; nach R. B. SETLOW, 1964)

Während bei allen bisher erwähnten Substraten die kompetitive Hemmung durch Präinkubation des Inhibitors mit dem Enzym im Licht wieder aufgehoben werden kann, macht UV-bestrahlte, Bromuracil (BU) enthaltende DNS eine Ausnahme. Als Inhibitor bindet sie das photoreaktivierende Enzym, jedoch ist die Hemmwirkung nach Belichtung nicht aufgehoben. Wahrscheinlich ist das Enzym irreversibel gebunden oder wird bei der Wechselwirkung mit dem Substrat inaktiviert (RUPERT, 1961; 1966).

Mit Ausnahme von BU-haltiger DNS sind bei allen Verbindungen mit Inhibitorwirkung Pyrimidin-Dimere auch nachgewiesen worden. Auch die bei Bestrahlung von DNS gebildeten „gemischten" Dimere C=T und U=T (letzteres entstanden durch Desaminierung von C=T), sowie U=U (entstanden aus C=C) werden durch das photoreaktivierende Enzym wieder gespalten. Dabei nimmt die Geschwindigkeit der Spaltung in der Reihenfolge T=T, C=T und U=T, C=C und U=U ab (R. B. SETLOW und CARRIER, 1966).

4. Nicht-photoreaktivierbare Schäden

Die Wirksamkeit, mit der UV-Schäden in der DNS durch PR eliminiert werden können, ist von vielen Faktoren (Wellenlänge des inaktivierenden UV-Lichtes, Basenzusammensetzung der DNS, Bedingungen der PR etc.) abhängig. Jedoch wird selbst unter optimalen Bedingungen nie eine quantitative Elimination von Letalschäden erreicht, gleichgültig ob es sich dabei um die PR von Zellen und Viren oder um die *in vitro* PR von Transformations- bzw. Virus-DNS handelt. Die Natur dieser irreparablen Schäden ist unbekannt.

Nach maximaler PR von UV-bestrahlter Transformations-DNS sind keine Thymin-Dimere mehr nachweisbar (J. K. Setlow, 1966a). Es ist hingegen noch nicht bekannt, ob auch alle Cytosin-Dimere wieder gespalten werden. Somit bleibt es noch offen, ob irreparable Schäden durch C=C verursacht werden können. Eine unvollständige Eliminierung der in einem DNS-Molekül vorhandenen C=C erscheint nicht ausgeschlossen, falls die Basensequenz in der Nachbarschaft eines Dimeren die Wechselwirkung des photoreaktivierenden Enzyms mit der DNS an dieser Stelle beeinflußt. Auch kann innerhalb einer kurzen Zeitspanne zwischen UV-Bestrahlung und PR schon eine Desaminierung eines Cytosin-Dimeren eingetreten sein. Dann würde selbst bei sonst vollständiger Spaltung aller dimeren Verbindungen durch die Substitution eines Cytosins durch ein Uracil in der DNS eine irreparable Veränderung resultieren. Einen Hinweis dafür, daß eine Cytosinveränderung für die Entstehung eines solchen Schadens in Frage kommen könnte, bietet das Aktionsspektrum für die nicht-photoreaktivierbaren Schäden. Es ist dem UV-Spektrum des Desoxycytidins sehr ähnlich, d.h. einem Maximum (Minimum) in der UV-Absorption entspricht ein hoher (geringer) Anteil an nicht-photoreaktivierbaren Schäden in der DNS (J. K. Setlow, 1963).

5. Übereinstimmung von in vitro- und in vivo-Photoreaktivierung

Der Nachweis, daß der PR von UV-geschädigten Zellen derselbe enzymatische Reaktivierungsmechanismus zugrundeliegt, wie er *in vitro* beobachtet wird, kann erst dann als gesichert gelten, wenn die Ergebnisse der *in vitro*-Photoreaktivierung mit denen der *in vivo*-PR in Übereinstimmung gebracht werden können. Für Bakterienzellen (und Viren) trifft es zu, daß unter Bedingungen, unter denen indirekte Photoreaktivierungsprozesse ausgeschaltet bzw. unterdrückt sind, die beobachtete PR auf die Wirkung eines photoreaktivierenden Enzyms zurückzuführen ist.

Eine Übereinstimmung zwischen der PR *in vivo* und der *in vitro* besteht schon in bezug auf die Strahlendosen und Wellenlängen, die bei der Inaktivierung und Reaktivierung angewendet werden, sowie in dem Grad der Wirksamkeit der PR (Rupert, 1960a; 1964a). Die Mitwirkung eines photoreaktivierenden Enzyms bei der PR einer Zelle wird auch schon dadurch verdeutlicht, daß photoreaktivierbare Organismen (*E. coli*, Hefe) einen enzymatisch aktiven Extrakt liefern, nicht-photoreaktivierbare Zellen (*E. coli* phr^-, Haemophilus influenzae) hingegen keinen (Rupert, 1965; Goodgal et al., 1957).

Ein direkter Nachweis, daß während der PR einer Zelle der photoreaktivierbare Schaden aus ihrer DNS entfernt wird, ist mit UV-bestrahlten *E. coli*-Bakterien durchgeführt worden. Diese erlangen durch die PR ihre Koloniebildungsfähigkeit zurück. Die damit verbundene Photoelimination der UV-Schäden wird begleitet von dem Verlust der hitzestabilisierenden Wirkung der Zell-DNS gegenüber dem photoreaktivierenden Enzym aus Hefe (Abschnitt IV, 2, 3) (RUPERT, 1965). Aus den UV-bestrahlten Bakterien werden nach Belichten (PR) oder nach Inkubation im Dunkeln die Nucleinsäuren extrahiert und mit dem Enzym zusammen erhitzt (2 min, 65° C). Anschließend wird die *verbliebene* Enzymaktivität im *in vitro*-System mit UV-bestrahlter Transformations-DNS bestimmt. Eine geringe Enzymaktivität resultiert, wenn die DNS aus der Zelle das Enzym nicht stabilisierte. In diesem Falle besaß die DNS keine UV-Schäden mehr. Ist die verbliebene enzymatische Aktivität jedoch noch hoch, muß sie zuvor stabilisiert gewesen sein, d.h. die Zell-DNS enthielt noch UV-Schäden. Auf diese Weise konnte gezeigt werden, daß UV-inaktivierte Wildtypzellen von *E. coli* (wie auch von *E. coli* B/r) während der PR ihrer Koloniebildungsfähigkeit die UV-Schäden in ihrer DNS eliminieren. Hingegen bleibt in UV-bestrahlten *E. coli* phr^--Zellen bei Belichtung der UV-Schaden in der DNS erhalten.

Die Vergleichbarkeit von *in vitro* und *in vivo*-PR wird auch durch den Nachweis einer intracellulären kompetitiven Hemmung der PR verdeutlicht (HARM und HILLEBRANDT, 1963). Die Geschwindigkeit der PR von gering UV-inaktivierten T4-Phagen in *E. coli* B wird herabgesetzt, wenn die Zellen gleichzeitig mit stark UV-bestrahlten Phagen desselben Genotyps infiziert werden. (Bei den letzteren ist die UV-Dosis so groß, daß ihre Überlebensrate selbst nach maximaler PR sehr klein, d.h. der nicht-photoreaktivierbare Schaden sehr groß ist.) Keine Verringerung der PR-Geschwindigkeit wird beobachtet, wenn vor der Infektion mit den leicht bestrahlten Phagen die stark inaktivierten Phagen maximal photoreaktiviert werden. Die hier beobachtete kompetitive Hemmung des *E. coli*-Enzyms konnte — im Gegensatz zu der des Enzyms aus Hefe — *in vitro* noch nicht nachgewiesen werden (RUPERT, 1964a).

Auch die (irreversible) Hemmung des photoreaktivierenden Enzyms durch BU-haltige DNS (Abschnitt IV, 3) hat ihre Parallele im Zellgeschehen. *E. coli*-Zellen, in denen etwa 35 % des Thymins durch BU ersetzt sind, lassen sich nach UV-Bestrahlung ebensowenig photoreaktivieren wie sich andererseits das photoreaktivierende Enzym von demselben DNS-Material bei Belichtung *in vitro* loslösen läßt (RUPERT, 1961). BU-haltige Bakteriophagen sind nach UV-Bestrahlung ebenfalls nicht photoreaktivierbar (STAHL et al., 1961).

6. Die Bedeutung der Photoreaktivierung in der Natur

Die im Laboratorium beobachtete PR von biologischen Systemen ist zu verstehen als eine einer spezifischen Schädigung des Systems entgegenwirkende Reaktion desselben. Die Vermutung liegt daher nahe, daß in der natürlichen Umgebung die Notwendigkeit besteht — oder zumindest einmal bestanden hat — einen solchen Reaktivierungsmechanismus zu entwickeln und zu unterhalten, um die Überlebenschance eines Organismus zu erhöhen. Wenn diese Hypothese richtig ist, dann müßten in der Natur dieselben Möglichkeiten zu spezifischen Schädigungen, wie sie durch UV-Bestrahlung im Laboratorium hervorgerufen werden, einmal vorhanden gewesen bzw. noch gegeben sein.

Wie eingangs erwähnt (II, 1), werden in den Laboratorien UV-Strahlenquellen benutzt, die hauptsächlich Licht der Wellenlänge 2537 Å emittieren. Die Strahlung enthält daneben auch noch einen geringen Anteil von noch kurzwelligerem Licht (1850 Å). Bestrahlungsversuche mit UV-Licht verschiedener Wellenlängen im Bereich von 2000—3000 Å haben gezeigt, daß die Wirkung des längerwelligen Lichtes (nahe 3000 Å) der des 2537 Å-Lichtes ähnlich ist (RUPERT, 1964a). Deshalb sind die bei 2537 Å-Bestrahlung beobachteten photobiologischen und -chemischen Prozesse auch für die im Wellenlängenbereich um 3000 Å sich abspielenden Vorgänge relevant. Nur die Letzteren spielen in der Natur eine Rolle, denn wie aus der Analyse des in der Natur vorkommenden UV-Lichtes hinsichtlich seiner Wellenlängenzusammensetzung hervorgeht, wird die Erdoberfläche nur von Strahlen der Wellenlänge bis herab zu 2900 Å erreicht. Der kurzwelligere Anteil des UV-Lichtes der Sonne wird durch eine Ozonschicht in der Atmosphäre, die selbst durch Reaktion des UV-Lichtes mit Sauerstoffmolekülen entsteht, absorbiert (SANDERSON und HULBURT, 1955). Die UV-Strahlung im Wellenlängenbereich von 2900 bis 3000 Å ist intensiv genug, um innerhalb eines Zeitraums von einigen Minuten bis zu einer Stunde eine Inaktivierung eines biologischen Systems beobachten zu können (RUPERT, 1964a).

Transformations-DNS, in UV-Licht durchlässigen Quarzgefäßen der Wirkung von Sonnenlicht ausgesetzt, wird im Laufe von Minuten inaktiviert. Der Zusatz von photoreaktivierendem Enzym aus Hefe bewirkt eine rasche, wenn auch nicht vollständige Reaktivierung (Abb. 7A). Dies ist möglich, weil neben den inaktivierenden UV-Strahlen gleichzeitig die photoreaktivierende Strahlung des Sonnenlichtes zugegen ist. Ist das Enzym vom Beginn der Sonnenbestrahlung an dabei, so wird die Inaktivierungsgeschwindigkeit der Transformations-DNS drastisch herabgesetzt (Abb. 7B) (RUPERT, 1960b). Dem Sonnenlicht ausgesetzte Bakterienzellen verhalten sich ähnlich (HARM, 1966). Die Fähigkeit zur Koloniebildung der *E. coli* phr$^-$-Mutante wird durch Sonnenbestrahlung viel rascher inaktiviert als die des Wildtyps *E. coli* phr$^+$. Parallel damit lassen sich in der DNS der phr$^-$-Mutante, nicht aber in der des Wildtyps photoreaktivierbare UV-Schäden nachweisen (RUPERT und HARM, 1966). (Diesem Nachweis liegt die im Abschnitt IV, 5 beschriebene Wechselwirkung der isolierten Bakterien-DNS mit dem photoreaktivierenden Enzym zugrunde.) Allerdings eliminiert das photoreaktivierende System in Zellen die durch Sonnenbestrahlung hervorgerufenen Letalschäden weniger effizient als die durch 2537 Å-UV-Licht erzeugten (HARM, 1966). Vermutlich entstehen durch die Strahlen des langwelligen UV-Lichtes auch nicht-reaktivierbare Letalschäden in anderen Zellbestandteilen.

Bedenkt man, daß sich die Ozonschicht in der Atmosphäre, welche die intensive kurzwellige UV-Strahlung absorbiert, erst im Laufe der Erdentstehung ausgebildet hat, so müssen die Organismen früher viel stärker den inaktivierenden Strahlen ausgesetzt gewesen sein. Es sollte daher eine evolutionsgeschichtlich sehr frühe Ausbildung von photoreaktivierenden Systemen stattgefunden haben. In Übereinstimmung damit steht das weitverbreitete Vorkommen lichtabhängiger Reaktivierungsmechanismen in der Natur. Betrachtet man aber die wenigen Objekte, bei denen bisher das Vorkommen

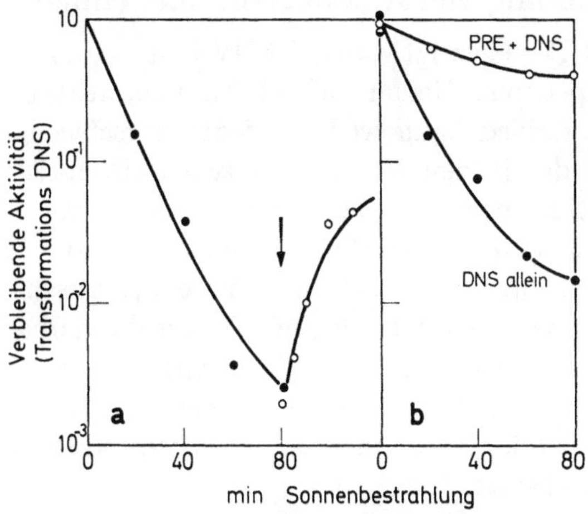

Abb. 7a u. b. Inaktivierung von Transformations-DNS durch Sonnenlicht und ihre Reaktivierung durch das photoreaktivierende Enzym aus Hefe. Transformations-DNS (Haemophilus influenzae) wird in Quarzgefäßen dem Sonnenlicht ausgesetzt. -●- DNS allein; -○- DNS mit photoreaktivierendem Enzym (PRE) aus Hefe. a Zusatz von Enzym (Pfeil) nach einer vorausgegangenen Inaktivierungsperiode. b Das Enzym ist vom Beginn der Sonnenbestrahlung an dabei. (Modifiziert nach RUPERT, 1960b)

eines photoreaktivierenden Enzyms nachgewiesen wurde, so ist die Notwendigkeit des Vorhandenseins eines Enzyms aufgrund der augenblicklichen biologischen Umgebung oftmals nicht gegeben.

Die Nützlichkeit der Existenz eines photoreaktivierenden Systems in Hefezellen ist offenkundig, denn Hefen sind bei ihrem Vorkommen an der offenen Luft gewöhnlich einer intensiven Sonnenbestrahlung ausgesetzt. Auch das Fehlen des Enzyms in Haemophilus influenzae, einem Mikroorganismus, der im nasopharyngealen Trakt anzutreffen ist, erscheint verständlich. Schwierigkeiten bereitet es aber schon, einzusehen, warum Enterobakterien wie *E. coli* ein photoreaktivierendes Enzym besitzen, und das Vorhandensein eines solchen Enzyms in der Leber eines Vertebraten (COOK und MCGRATH, 1967) kann sicher nichts mehr mit der Photoreaktivierung UV-Strahlen-induzierter Dimere in der DNS zu tun haben. Vielleicht war der Bedarf an Photoreaktivierungssystemen zu Beginn der Evolution so groß, daß sich zu dessen Synthesen ein entsprechend umfangreicher Genapparat entwickelt hatte, der im Verlauf der Evolution zwar stark verkümmert ist, aber in Restbeständen immer noch in vielen Organismen anzutreffen ist. Oder aber wir beobachten in den Photoreaktivierungsprozessen gar nicht die eigentlichen physiologischen Funktionen, die normalerweise den daran beteiligten Enzymen in der Zelle zukommen. Beide Erklärungsmöglichkeiten sind aber von dem Gesichtspunkt her, daß die bisher bekannten photoreaktivierenden Enzyme mit großer Effizienz arbeiten, d.h. bis zu 90% der durch UV-Licht induzierten Letalschäden eliminieren können, wenig zufriedenstellend.

V. Photoreaktivierung von Ribonucleinsäure enthaltenden Viren

Wie in Abschnitt IV, 3 gezeigt wurde, ist UV-bestrahlte RNS im Gegensatz zur DNS kein kompetitiver Hemmstoff für das photoreaktivierende Enzym aus Hefe. Der Unterschied kann weder auf den verschiedenen makromolekularen Strukturen der beiden Nucleinsäuretypen (Doppelstrang bei DNS, Einzelstrang bei RNS), noch auf den verschiedenen Dimertypen (Thymin-Dimere bei DNS, Uracil-Dimere bei RNS) beruhen, denn sowohl UV-bestrahlte einsträngige *DNS* als auch *Uracil*-Dimere in enzymatisch synthetisierten Polymeren aus Desoxyribonucleotiden sind photoreaktivierbar bzw. spaltbar (RUPERT, 1964a; SETLOW et al., 1965). Die Annahme ist daher berechtigt, daß das photoreaktivierende Enzym aus Hefe, ebenso wie viele andere Enzyme des Nucleinsäurestoffwechsels, zwischen Desoxyribose und Ribose enthaltenden Nucleinsäuren unterscheidet.

Es darf aber nicht übersehen werden, daß UV bestrahlte Ribonucleinsäuren — im Gegensatz zu Desoxyribonucleinsäuren — neben den dimerisierten Pyrimidinbasen auch Hydrate dieser Basen enthalten (Abschnitt II, 2). Die Geschwindigkeit der Hydratisierung von Uracilbasen ist zudem — zumindest in der Polyuridylsäure — größer als die der Dimerisierung (PEARSON und JOHNS, 1966). Wahrscheinlich werden auch in RNS-Molekülen, die alle vier Basen (Adenin, Guanin, Cytosin und Uracil) enthalten, pro UV-Dosis mehr Hydrate als Dimere gebildet, weil *jede* Pyrimidinbase ein Hydrat bilden kann, während sie nur dann dimerisieren kann, wenn in Nachbarschaft eine zweite Pyrimidinbase vorhanden ist. Ein nach UV-Bestrahlung überwiegend Hydrate enthaltendes RNS-Molekül könnte aber die Wechselwirkung des photoreaktivierenden Enzyms mit den in demselben Molekül vorhandenen Dimeren beeinträchtigen. Wenn dies auch nicht sehr wahrscheinlich ist, so steht doch der Versuch noch aus, eine ausschließlich Dimere enthaltende RNS daraufhin zu prüfen, ob sie als Substrat für das Enzym in Frage kommt.

Ganz unabhängig von den negativen Ergebnissen mit dem photoreaktivierenden Enzym aus Hefe sind andere biologische Systeme daraufhin getestet worden, ob sie die Kapazität zur Photoreaktivierung von RNS besitzen. Untersucht wurden RNS-Bakteriophagen in Bakterienzellen und die ebenfalls RNS enthaltenden Pflanzenviren. Nur bei den Letzteren, und auch hier nicht bei allen, ist eine Photoreaktivierung eindeutig beobachtet worden. Jedoch ist in den entsprechenden Systemen der Nachweis eines photoreaktivierenden Enzyms bisher noch nicht gelungen.

1. RNS-Bakteriophagen

Mit einer einzigen Ausnahme sind alle bisher untersuchten RNS-Bakteriophagen, deren Wirtsbakterien entweder *E. coli* oder Pseudomonas aeruginosa sind, nicht photoreaktivierbar (WINKLER, 1964; ZINDER, 1965; RAUTH, 1965. WERBIN et al., 1967). Die Ausnahme macht der Phage f2, jedoch ist es auch hier noch umstritten, ob eine Photoreaktivierung wirklich vorhanden ist. Nach extracellulärer UV-Bestrahlung des intakten Virus und nachfolgender Infektion der Bakterienzelle ist keine PR beobachtet worden, jedoch tritt ein geringer lichtabhängiger Reaktivierungseffekt auf, wenn die RNS des Phagen in Lösung bestrahlt wird und anschließend mit ihr Protoplasten von *E. coli* infiziert werden (WERBIN et al., 1967). Etwa 17% der UV-Schäden werden eliminiert. Man muß dabei bedenken, daß bei den *E. coli*-Phagen, die einsträngige *DNS* enthalten (ΦX174, S13, fd), auch nicht mehr als 20—30% aller Schäden photoreaktiviert werden (HARM, 1967). Die Tatsache, daß nach Infektion intakter Zellen mit UV-bestrahlten f2-Phagen keine PR erfolgt, könnte auf der beobachteten Beeinträchtigung des Penetrationsvermögens strahlengeschädigter RNS-Moleküle durch die Zellwand beruhen (WERBIN et al., 1968). Diese Erklärungsmöglichkeit läßt sich aber für die Nicht-photoreaktivierbarkeit eines anderen RNS-haltigen Virus, dem Bakteriophagen fr, nicht heranziehen, denn bei ihm wurde auch nach UV-Bestrahlung von *phageninfizierten Zellen* keine PR beobachtet (WINKLER, 1964).

Die Nicht-photoreaktivierbarkeit des Phagen fr ist insofern von Bedeutung, als seine Nucleinsäure etwa denselben Anteil an Pyrimidinbasen enthält wie die des Kartoffel X-Virus. Eine UV-Inaktivierung dürfte deshalb in beiden Virusarten durch annähernd dieselben photochemischen Reaktionen verursacht werden (WINKLER, 1964). Unter den photoreaktivierbaren phytopathogenen Viren zeigt aber gerade das Kartoffel-X-Virus die größte Reaktivierbarkeit (BAWDEN und KLECZKOWSKI, 1955). Man wird daher annehmen müssen, daß das photoreaktivierende System in der Pflanzenzelle von dem in der Bakterienzelle verschieden ist.

2. RNS-Pflanzenviren

Bei einer Reihe von Pflanzenviren ist die Überlebensrate nach UV-Bestrahlung verschieden, je nachdem ob die Pflanzen nach der Infektion dem Licht ausgesetzt, oder erst für längere Zeit im Dunkeln gehalten werden (BAWDEN und KLECZKOWSKI, 1955). In beiden Fällen folgt die Inaktivierung einer Eintreffer-Reaktion nach der Exponentialfunktion:

$$S/S_0 = e^{-k_{d,l}\cdot t} \quad \text{oder} \quad \ln S/S_0 = -k_{d,l}\cdot t$$

wobei S/S_0 die Fraktion der Überlebenden nach t Minuten Bestrahlung wiedergibt und k_d oder k_l Konstanten darstellen, die für die UV-Empfindlichkeit des Virus (bzw. der Virusnucleinsäure) bei Infektion im Dunkeln (k_d) oder im Licht (k_l) charakteristisch sind. Wenn keine PR stattgefunden hat, ist $k_d = k_l = 1{,}0$, bei vorhandener PR ist $k_d > k_l$ und das Verhältnis k_d/k_l ist dabei umso größer, je effektiver die PR ist.

In Tabelle 2 sind die Quotienten k_d/k_l für die bisher untersuchten phytopathogenen Viren und ihre infektiösen Ribonucleinsäuren aufgeführt. In dieser Darstellung fällt auf, daß nur die beiden Virusarten, die aus starren Stäbchen bestehen, keine PR zeigen, während die Viren mit polyedrischer Gestalt und diejenigen mit flexibler Stäbchenform unterschiedliche Grade in ihrer Reaktivierbarkeit aufweisen. Die flexibel-stäbchenförmigen Viren, besonders das Kartoffel-X-Virus, werden am effektivsten photoreaktiviert. Wie am Beispiel des Tabak-Nekrose-Virus gezeigt wird, ist bei den photoreaktivierbaren Viren das Ausmaß der PR offenbar dasselbe, gleichgültig ob das intakte Virus oder die Virus-RNS bestrahlt werden (KASSANIS und KLECZKOWSKI, 1965). Eine PR ist auch beim Tabakmosaikvirus und Tabak-Rattle-Virus zu beobachten, wenn nicht die intakten Viren, sondern die von der Proteinhülle befreiten Virusnucleinsäuren in Lösung bestrahlt werden (BAWDEN und KLECZKOWSKI, 1959; HARRISON und NIXON, 1959).

Die Photoreaktivierbarkeit intakter Pflanzenviren ist abhängig von der Art der Nucleoproteidstruktur der einzelnen Virusarten. Bei den stäbchenförmigen Viren liegt die RNS in einer Helixstruktur vor. Je nachdem, wie stark dabei die Bindung zwischen der RNS und dem Hüllprotein ist, sind die Stäbchen starr oder flexibel. Bei starker Bindung (starre Stäbchen), nicht jedoch bei schwacher Bindung (flexible Stäbchen) verhindert das Protein die Bildung photoreaktivierbarer Schäden in der RNS. Entsprechend sind alle polyedrischen Viren, bei denen keine oder nur eine schwache Bindung zwischen Nucleinsäure und Protein existiert, photoreaktivierbar. Innerhalb der Gruppe der photoreaktivierbaren Viren ist das *Ausmaß* der Reaktivierung wiederum sehr verschieden. Hierfür müssen andere Faktoren wie die unterschiedliche Basenzusammensetzung und -sequenz der Ribonucleinsäuren, sowie die PR-Kapazität der verschiedenen Pflanzenarten verantwortlich gemacht werden.

Bei den Stämmen U1 und U2 des Tabakmosaikvirus ist der Einfluß der Proteinhülle auf die UV-Inaktivierung und die PR der Nucleinsäure eingehender untersucht worden. Die Ribonucleinsäuren beider Stämme besitzen dieselbe UV-Empfindlichkeit, jedoch ist das intakte U1-Virus etwa um den Faktor 6 resistenter gegen UV-Strahlen als das intakte U2-Virus (SIEGEL et al., 1956). Demnach schützt das Protein vom Stamm U1, welches sich von dem Protein vom Stamm U2 in mindestens 33 Aminosäure-Positionen unterscheidet (WITTMANN-LIEBOLD und WITTMANN, 1967), die Nucleinsäure stärker vor der Inaktivierung durch UV-Strahlen. Daß die abschirmende Wirkung allein dem Virusprotein zukommt, erkennt man auch daran, daß die durch Rekonstitution gewonnenen hybriden Viren U1 RNS — U2 Protein und U2 RNS — U1 Protein dieselben UV-Empfindlichkeiten besitzen wie der native Stamm, von dem das *Protein* stammt (STRETTER und GORDON, 1967).

Wenn nach UV-Bestrahlung des Tabakmosaikvirus (Stamm U1) die noch verbliebenen Infektiositäten sowohl von dem intakten Virus wie auch von der aus dem bestrahlten Virus isolierten RNS bestimmt werden, so findet man, daß die Inaktivierungsgeschwindigkeiten von Virus und Virus-RNS identisch

Tabelle 2. *Photoreaktivierung von Pflanzenviren*

Virus (Virus-RNS)	Gestalt des Virus	k_d/k_l	Literatur
Tabakmosaikvirus[a] (Tobacco Mosaic)	Stäbchen (starr)	1,0	BAWDEN und KLECZKOWSKI (1955)
Tabakmosaikvirus-RNS	—	2,0[b]	BAWDEN und KLECZKOWSKI (1959), RUSHIZKY et al. (1960)
Tabak-Rattle-Virus (Tobacco Rattle)	Stäbchen (starr)	1,0	CADMAN und HARRISON (1959)
Tabak-Rattle-Virus-RNS	—	1,4—1,9	HARRISON und NIXON (1959)
Bushystunt-Virus der Tomate (Tomato Bushy Stunt)	polyedrisch	1,2	BAWDEN und KLECZKOWSKI (1955)
Tabak-Nekrose-Virus (Tobacco Necrosis)	polyedrisch	1,35	KASSANIS und KLECZKOWSKI (1965)
Tabak-Nekrose-Virus-RNS	—	1,35	KASSANIS und KLECZKOWSKI (1965)
Gurkenmosaikvirus (Cucumber Mosaic)	polyedrisch	1,5	BAWDEN und KLECZKOWSKI (1955)
Gelbmosaikvirus des Klees (Clover Yellow Mosaic)	Stäbchen (flexibel)	1,8[c]	CHESSIN (1965)
Tabakringfleckenvirus (Tobacco Ringspot)	polyedrisch	1,9	BAWDEN und KLECZKOWSKI (1955)
Schwarzringfleckenvirus des Kohls (Cabbage-Black Ringspot)	Stäbchen (flexibel)	2,0	BAWDEN und KLECZKOWSKI (1955)
Kartoffel X-Virus (Potato X)	Stäbchen (flexibel)	3,1	BAWDEN und KLECZKOWSKI (1955)

[a] 5 verschiedene Stämme, unter ihnen U1 und U2, wurden untersucht.

[b] Gemessen im Wellenlängebereich 2300—2800 Å (RUSHIZKY et al., 1960).

[c] Gemessen bei hoher UV-Dosis, da bei niederer Dosis die PR durch einen Inhibitoreffekt des Lichtes auf die Virusinfektion maskiert wird.

sind (GODDARD et al., 1966). Alle bei der Bestrahlung des intakten Virus erzeugten Letalschäden müssen demnach in dessen RNS auftreten. Keiner dieser Schäden ist jedoch photoreaktivierbar, ganz im Gegensatz zu den photoreaktivierbaren Schäden, die bei Bestrahlung der Virus-*RNS* zu finden sind (BAWDEN und KLECZKOWSKI, 1959). Die eigenartige Wirkungsweise der Virusproteinhülle besteht demnach darin, daß sie die Bildung photoreaktivierbarer Schäden unterdrückt, die RNS aber dennoch nicht so gut abschirmt, daß nicht doch letale Veränderungen in ihr auftreten.

Neuere Untersuchungen haben gezeigt, daß bei UV-Bestrahlung von Tabakmosaikvirus Proteinuntereinheiten an die RNS gebunden werden. Etwa eine Untereinheit wird pro Letalschaden an der RNS fixiert (GODDARD et al., 1966) Dieses Protein läßt sich nicht durch warmes Natriumdodecylsulfat, wohl aber durch Phenol von der Nucleinsäure wieder entfernen. Es ist demnach nicht durch eine kovalente Bindung mit der Nucleinsäure verknüpft. Auffallend ist, daß bestrahlte Tabakmosaikvirus-RNS nach der Rekonstitution zum intakten Virus nicht mehr photoreaktivierbar ist (SMALL und GORDON, 1967). Wahrscheinlich verhindern die UV-Schäden in der Nucleinsäure die normalerweise erfolgende Dissoziation des Nucleoproteids in der Zelle. Die Verhinderung der PR ist aber nicht irreversibel, denn die RNS kann wieder aus dem rekonstituierten, nicht-photoreaktivierbaren Virus extrahiert werden und ist dann wieder photoreaktivierbar.

Es ist noch nicht bekannt, welche Art der chemischen Veränderung in den Ribonucleinsäuren der Pflanzenviren photoreaktiviert werden kann. Vergleicht man bei der TMV-RNS die Dosisabhängigkeit der Bildung von Uracil-Dimeren mit der der Inaktivierung, so findet man (unter sehr verschiedenen Strahlungsbedingungen) pro letalen „Treffer" weniger als ein Uracil-Dimeres pro RNS-Molekül. Die Dimerisierung kann also nicht — zumindest nicht allein — für die letale Veränderung verantwortlich gemacht werden. Wahrscheinlich kommt auch der Hydratbildung der Pyrimidine eine biologische Bedeutung zu (MERRIAM und GORDON, 1967). Untersuchungen über den Verbleib der Photoprodukte, wie sie für die DNS in Bakterienzellen durchgeführt wurden (s. IV, 5), sind für die RNS-haltigen Pflanzenviren noch nicht bekanntgeworden.

Literatur

BAWDEN, F. C., and A. KLECZKOWSKI: Studies on the ability of light to counteract the inactivating action of ultraviolet radiation on plant viruses. J. gen. Microbiol. **13**, 370—382 (1955).

— — Photoreactivation of nucleic acid from tobacco mosaic virus. Nature (Lond.) **183**, 503—504 (1959).

BEUKERS, R.: The effect of proflavine on u.v.-induced dimerization of thymine in DNA. Photochem. Photobiol. **4**, 935—937 (1965).

—, and W. BERENDS: Isolation and identification of the irradiation product of thymine. Biochim. biophys. Acta (Amst.) **41**, 550—551 (1960).

BOYCE, R. P., and P. HOWARD-FLANDERS: Release of ultraviolet light-induced thymine dimers from DNA in E. coli K-12. Proc. nat. Acad. Sci. (Wash.) **51**, 293—300 (1964).

CADMAN, C. H., and B. D. HARRISON: Studies on the properties of soil-borne viruses of the tobacco-rattle type occurring in Scotland. Ann. appl. Biol. **47**, 542—556 (1959).

CHESSIN, M.: Photoreactivation of clover yellow mosaic virus. Photochem. Photobiol. **4**, 709—711 (1965).

CLEAVER, J. E.: Photoreactivation: A radiation repair mechanism absent from mammalian cells. Biochem. biophys. Res. Commun. **24**, 569—576 (1966).

COOK, J. S.: Direct demonstration of the monomerization of thymine-containing dimers in u.v.-irradiated DNA by yeast photoreactivating enzyme and light. Photochem. Photobiol. **6**, 97—101 (1967).

COOK, J. S., and J. R. MCGRATH: Photoreactivating-enzyme activity in metazoa. Proc. nat. Acad. Sci. (Wash.) **58**, 1359—1365 (1967).

DAVID, C. N.: UV inactivation and thymine dimerization in bacteriophage ΦX. Z. Vererbungsl. **95**, 318—325 (1964).

DOUDNEY, C. O.: I. Ultraviolet light effects on the bacterial cell. Current Topics Microbiol. Immunol. **46**, 116—175 (1968).

DULBECCO, R.: Reactivation of ultra-violet-inactivated bacteriophage by visible light. Nature (Lond.) **163**, 949—950 (1949).

FRUTON, J. S., and S. SIMMONDS: In: General biochemistry. New York: John Wiley & Sons 1958.

GAREN, A., and N. D. ZINDER: Radiological evidence for partial genetic homology between bacteriophage and host bacteria. Virology **1**, 347—376 (1955).

GODDARD, J., D. STREETER, C. WEBER, and M. P. GORDON: Studies on the inactivation of tobacco mosaic virus by ultraviolet light. Photochem. Photobiol. **5**, 213—222 (1966).

GOODGAL, S. H., C. S. RUPERT, and R. M. HERRIOTT: Photoreactivation of hemophilus influenzae transforming factor for streptomycin resistence by an extract of escherichia coli B. In: The chemical basis of heredity. Baltimore: Johns Hopkins Press 1957.

GORDON, M. P., and M. STAEHELIN: Studies on the incorporation of 5-fluorouracil into a virus nucleic acid. Biochim. biophys. Acta (Amst.) **36**, 351—361 (1959).

GRIFFITH, J. D., u. R. B. SETLOW: Unveröffentlichte Versuche. Zit. in SETLOW et al. 1965.

HANAWALT, P. C.: The U.V. sensitivity of bacteria: its relation to the DNA replication cycle. Photochem. Photobiol. **5**, 1—12 (1966).

—, and R. H. HAYNES: Repair replication of DNA in bacteria: irrelevance of chemical nature of base defect. Biochem. biophys. Res. Commun. **19**, 462—467 (1965).

HARIHARAN, P. V., and H. E. JOHNS: Rate constants for the dehydration of single and double hydrates of cytidylyl (3'-5')-cytidine. Photochem. Photobiol. **7**, 239—259 (1968).

HARM, W.: Repair effects in phage and bacteria exposed to sunlight. Radiat. Res. Suppl. **6**, 215 (1966).

— Zit. in WERBIN et al. 1967.

—, and B. HILLEBRANDT: A non-photoreactivable mutant of E. coli B. Photochem. Photobiol. **1**, 271—272 (1962).

— — Kompetitive Hemmung der Photo-Reaktivierung von uv-inaktivierten T4-Phagen durch stark uv-bestrahlte Phagen-DNA. Z. Naturforsch. **18**b, 294—300 (1963).

—, and C. S. RUPERT: Infection of transformable cells of haemophilus influenzae by bacteriophage and bacteriophage DNA. Z. Vererbungsl. **94**, 336—348 (1963).

HARRISON, B. D., and H. L. NIXON: Some properties of infective preparations made by disrupting tobacco rattle virus with phenol. J. gen. Microbiol. **21**, 591—599 (1959).

HOWARD-FLANDERS, P., and R. P. BOYCE: DNA repair and genetic recombination: Studies on mutants of Escherichia coli defective in these processes. Radiat. Res., Suppl. **6**, 156—184 (1966).

JAGGER, J.: Photoreactivation. Bact. Rev. **22**, 99—142 (1958).

— Photoprotection from ultraviolet killing in Escherichia coli B. Radiat. Res., Suppl. **13**, 521—539 (1960).

—, and R. S. STAFFORD: Evidence for two mechanisms of photoreactivation in Escherichia coli B. Biophys. J. **5**, 75—88 (1965a).

— — Evidence that indirect photoreactivation does not split thymine dimers. Abstract. Pacific Slope Biochemical Conference, University of Southern California, Los Angeles 1965b.

JAGGER, J., W. S. WISE, and R. S. STAFFORD: Delay in growth and division induced by near ultraviolet radiation in Escherichia coli B and its role in photoprotection and liquid holding recovery. Photochem. Photobiol. **3**, 11—24 (1964).

JOHNS, H. E., S. A. RAPAPORT, and M. DELBRÜCK: Photochemistry of thymine dimers. J. molec. Biol. **4**, 104—114 (1962).

KASSANIS, B., and A. KLECZKOWSKI: Inactivation of a strain of tobacco necrosis virus and of the RNA isolated from it, by ultraviolet radiation of different wave-lengths. Photochem. Photobiol. **4**, 209—214 (1965).

KELNER, A.: Effect of visible light on the recovery of streptomyces griseus conidia from ultra-violet irradiation injury. Proc. nat. Acad. Sci. (Wash.) **35**, 73—79 (1949).

LERMAN, L. S.: Structural considerations in the interaction of DNA and acridines. J. molec. Biol. **3**, 18—30 (1961).

MARMUR, J., and L. GROSSMAN: Ultraviolet light induced linking of deoxyribonucleic acid strands and its reversal by photoreactivating enzyme. Proc. nat. Acad. Sci. (Wash.) **47**, 778—787 (1961).

MCLAREN, A. D., and D. SHUGAR: Photochemistry of proteins and nucleic acids. Oxford: Pergamon Press 1964.

MERRIAM, V., and M. P. GORDON: Pyrimidine dimer formation in ultraviolet irradiated TMV-RNA. Photochem. Photobiol. **6**, 309—319 (1967).

MUHAMMED, A.: Studies on the yeast photoreactivating enzyme. J. biol. Chem. **241**, 516—523 (1966).

PEARSON, M. L., and H. E. JOHNS: Suppression of hydrate and dimer formation in ultraviolet-irradiated Poly (A + U) relative to Poly U. J. molec. Biol. **20**, 215—229 (1966).

— F. P. OTTENSMEYER, and H. E. JOHNS: Properties of an unusual photoproduct of U.V. irradiated thymidylyl-thymidine. Photochem. Photobiol. **4**, 739—747 (1965).

RAUTH, A. M.: The physical state of viral nucleic acid and the sensitivity of viruses to ultraviolet light. Biophys. J. **5**, 257—273 (1965).

REGAN, J. D., and J. S. COOK: Photoreactivation in an established vertebrate cell line. Proc. nat. Acad. Sci. (Wash.) **58**, 2274—2279 (1967).

RUPERT, C. S.: Photoreactivation of transforming DNA by an enzyme from Baker's yeast. J. gen. Physiol. **43**, 573—595 (1960a).

— In: The comparative effects of radiation. New York: John Wiley & Sons 1960b.

— Repair of ultraviolet damage in cellular DNA. J. cell. comp. Physiol. **58**, Suppl. 1, 57—68 (1961).

— Photoenzymatic repair of ultraviolet damage in DNA. I. Kinetics of the reaction. J. gen. Physiol. **45**, 703—724 (1962a).

— Photoenzymatic repair of ultraviolet damage in DNA. II. Formation of an enzyme substrate complex. J. gen. Physiol. **45**, 725—741 (1962b).

— Photoreactivation of ultraviolet damage. In: Photophysiology, vol. II. New York: Academic Press 1964a.

— Questions regarding the presumed role of thymine dimer in photoreactivable U.V. damage to DNA. Photochem. Photobiol. **3**, 399—403 (1964b).

— Relation of photoreactivation to photoenzymatic repair of DNA in Escherichia coli. Photochem. Photobiol. **4**, 271—275 (1965).

— 1966 Diskussionsbemerkung. Zit. in J. K. SETLOW 1966a.

— S. H. GOODGAL, and R. M. HERRIOTT: Photoreactivation in vitro of ultraviolet inactivated hemophilus influenzae transforming factor. J. gen. Physiol. **41**, 451—471 (1958).

— and W. HARM: Reactivation of photobiological damage. In: Advances in radiation biology, vol. II. New York: Academic Press 1966.

Rupp, W. D., and P. Howard-Flanders: Discontinuities in the DNA synthesized in an excision-defective strain of Escherichia coli following ultraviolet irradiation. J. molec. Biol. **31**, 291—304 (1968).

Rushizky, G. W., C. A. Knight, and A. D. McLaren: A comparison of the ultraviolet-light inactivation of infectious ribonucleic acid preparations from tobacco mosaic virus with those of the native and reconstituted virus. Virology **12**, 32—47 (1960).

Sanderson, J. A., and E. O. Hulburt: Sunlight as a source of radiation. In: Radiation biology, vol. II. New York: McGraw-Hill Book Co. 1955.

Sarin, P. S., and H. E. Johns: U.V. induced conformational changes in transfer RNA. Photochem. Photobiol. **7**, 203—210 (1968).

Schuster, H.: Photochemie von Ribonucleinsäuren. Z. Naturforsch. **19**b, 815—830 (1964).

Setlow, J. K.: The wavelength-dependent fraction of biological damage due to thymine dimers and to other types of lesion in ultraviolet-irradiated DNA. Photochem. Photobiol. **2**, 393—399 (1963).

— Photoreactivation. Radiat. Res., Suppl. **6**, 141—155 (1966a).

— The molecular basis of biological effects of ultraviolet radiation and photoreactivation. In: Current topics in radiation research, vol. II. Amsterdam: North-Holland Publ. Co. 1966b.

—, and M. E. Boling: The action spectrum of an in vitro DNA photoreactivation system. Photochem. Photobiol. **2**, 471—477 (1963).

— —, and F. J. Bollum: The chemical nature of photoreactivable lesions in DNA. Proc. nat. Acad. Sci. (Wash.) **53**, 1430—1436 (1965).

—, and F. J. Bollum: The minimum size of the substrate for yeast photoreactivating enzyme. Biochim. biophys. Acta (Amst.) **157**, 233—237 (1968).

—, and R. B. Setlow: Nature of the photoreactivable ultra-violet lesion in deoxyribonucleic acid. Nature (Lond.) **197**, 560—562 (1963).

— — Contribution of dimers containing cytosine to ultra-violet inactivation of transforming DNA. Nature (Lond.) **213**, 907—909 (1967).

Setlow, R. B.: Physical changes and mutagenesis. J. cell. comp. Physiol. **64**, Suppl. 1, 51—68 (1964).

— Repair of DNA. In: Regulation of nucleic acid and protein biosynthesis. Amsterdam: Elsevier Publ. Co. 1967.

—, and W. L. Carrier: The disappearance of thymine dimers from DNA: an error-correcting mechanism. Proc. nat. Acad. Sci. (Wash.) **51**, 226—231 (1964).

— — Pyrimidine dimers in ultraviolet-irradiated DNA's. J. molec. Biol. **17**, 237—254 (1966).

— — Formation and destruction of pyrimidine dimers in polynucleotides by ultraviolet irradiation in the presence of proflavine. Nature (Lond.) **213**, 906—907 (1967).

— —, and F. J. Bollum: Pyrimidine dimers in UV-irradiated Poly dI:dC. Proc. nat. Acad. Sci. (Wash.) **53**, 1111—1118 (1965).

—, and J. K. Setlow: Evidence that ultraviolet-induced dimers in DNA cause damage. Proc. nat. Acad. Sci. (Wash.) **48**, 1250—1257 (1962).

— P. A. Swenson, and W. L. Carrier: Thymine dimers and inhibition of DNA synthesis by ultraviolet irradiation of cells. Science **142**, 1464—1465 (1963).

Siegel, A., S. G. Wildman, and W. Ginoza: Sensitivity to ultra-violet light of infectious tobacco mosaic virus nucleic acid. Nature (Lond.) **178**, 1117—1118 (1956).

Small, G. D., and M. P. Gordon: Prevention of photoreactivation of tobacco mosaic virus-ribonucleic acid by reconstitution. Photochem. Photobiol. **6**, 303—308 (1967).

STAHL, F. W., J. M. CRASEMAN, L. OKUN, E. FOX, and C. LAIRD: Radiation-sensitivity of bacteriophage containing 5-bromo-deoxyuridine. Virology **13**, 98—104 (1961).

STRETTER, D. G., and M. P. GORDON: Ultraviolet photoinactivation studies on hybrid viruses obtained by cross-reconstitution of the protein and RNA components of U(1) and U(2) strains of TMV. Photochem. Photobiol. **6**, 413—421 (1967).

TERRY, C. E., and J. K. SETLOW: Photoreactivating enzyme from neurospora crassa. Photochem. Photobiol. **6**, 799—803 (1967).

WACKER, A.: Molecular mechanisms of radiation effects. In: Progress in nucleic acid research, vol. I. New York: Academic Press 1963.

WANG, S. Y., and R. ALCÁNTARA: The possible formation of dithymine peroxyde in irradiated DNA. Photochem. Photobiol. **4**, 477—481 (1965).

—, and A. J. VARGHESE: Cytosine-thymine addition product from DNA irradiated with ultraviolet light. Biochem. biophys. Res. Commun. **29**, 543—549 (1967).

WEATHERWAX, R. S.: Desensitization of Escherichia coli to ultraviolet light. J. Bact. **72**, 124—125 (1956).

WEINBLUM, D.: Characterization of the photodimers from DNA. Biochem. biophys. Res. Commun. **27**, 384—390 (1967).

—, and H. E. JOHNS: Isolation and properties of isomeric thymine dimers. Biochim. biophys. Acta (Amst.) **114**, 450—459 (1966).

WERBIN, H., and C. S. RUPERT: Presence of photoreactivating enzyme in blue-green algal cells. Photochem. Photobiol. **7**, 225—230 (1968).

— R. C. VALENTINE, O. HILDALGO-SALVATIERRA, and A. D. MCLAREN: Photobiology of RNA bacteriophages-II. U.V. irradiation of f2: effects on extracellular stages of infection and on early replication. Photochem. Photobiol. **7**, 253—261 (1968).

— —, and A. D. MCLAREN: Photobiology of RNA bacteriophages-I. Ultraviolet inactivation and photoreactivation studies. Photochem. Photobiol. **6**, 205—213 (1967).

WIERZCHOWSKI, K. L., and D. SHUGAR: Photochemistry of model oligo- and poly-nucleotides- IV. Hetero-oligonucleotides and high molecular weight single and twin-stranded polymer chains. Photochem. Photobiol. **1**, 21—36 (1962).

WINKLER, U.: Über die fehlende Photo- und Wirtszellenreaktivierbarkeit des UV-inaktivierten RNS-Phagen fr. Photochem. Photobiol. **3**, 37—43 (1964).

— H. E. JOHNS, and E. KELLENBERGER: Comparative study of some properties of bacteriophage T4D irradiated with monochromatic ultraviolet light. Virology **18**, 343—358 (1962).

WITTMANN-LIEBOLD, B., and H. G. WITTMANN: Coat proteins of strains of two RNA viruses: comparison of their amino acid sequences. Molec. Gen. Genetics **100**, 358—363 (1967).

WULFF, D. L., and C. S. RUPERT: Disappearance of thymine photodimer in ultraviolet irradiated DNA upon treatment with a photoreactivating enzyme from Baker's yeast. Biochem. biophys. Res. Commun. **7**, 237—240 (1962).

YAMANE, T., B. J. WYLUDA, and R. G. SHULMAN: Dihydrothymine from UV-irradiated DNA. Proc. nat. Acad. Sci. (Wash.) **58**, 439—442 (1967).

ZAMENHOF, S., B. REINER, R. DE GIOVANNI, and K. RICH: Introduction of unnatural pyrimidines into deoxyribonucleic acid of Escherichia coli. J. biol. Chem. **219**, 165—173 (1956).

ZINDER, N. D.: RNA phages. In: Ann. review microbiology, vol. 19. Palo Alto: Annual Reviews, Inc. 1965.

From the Section on Infectious Diseases, Perinatal Research Branch,
National Institute of Neurological Diseases and Blindness,
National Institutes of Health, Bethesda, Maryland 20014

Rubella Virus

LUIZ HORTA-BARBOSA, D. V. M., DAVID FUCCILLO, PH. D.,
and
JOHN L. SEVER, M. D., PH. D.

With 3 Figures

Table of Contents

Introduction

A variety of infectious agents are capable of producing perinatal diseases. Rubella is of particular importance because of the high frequency of the disease during epidemics and the wide spectrum of damage caused by the virus on the developing fetus. The virus is capable of producing chronic infection of the embryo which may persist after birth, together with congenital malformations, mental retardation, and severe tissue involvement of various organs.

It is our intention in this report to review the current status of knowledge on rubella virus and to summarize recent advances in understanding the mechanisms of infection as well as control of the disease.

Rubella as a Clinical Entity

Rubella was first recognized in 1815 when MATON described "a rash liable to be mistaken for scarlatina" (MATON). Over a century later GREGG noted a high incidence of congenital malformations, mostly cataracts, and reported the association of these abnormalities with rubella contracted in the first trimester of pregnancy (GREGG). Since the successful isolation of rubella virus in 1962 (PARKMAN, 1962; WELLER, 1962; SEVER, 1962a), several groups have reported on methods of detection of the virus.

It is well recognized that frequency and type of abnormality differ with the time of maternal infection. Almost all abnormalities are associated with infection in the first months of pregnancy. The frequency of production of abnormal children is approximately 50% in patients who contract rubella in the first month of pregnancy, 22% in the second month, and 6—8% in the third month (MICHAELS, LUNDSTROM). The major findings reported were heart lesions, cataracts, deafness, microcephaly, and mental retardation. Recent studies indicate that congenital deafness, mental and motor retardation may occur when rubella is contracted in the second or early third trimester of pregnancy (HARDY, in press). Chronic infection of the fetus usually persists in the child after birth for a period of months to years. The virus can be readily recovered from throat, urine, spinal fluid and most tissues. Children with congenital rubella also may show thrombocytopenia with petechiae and ecchymoses (50%), hepatosplenomegaly (40%), pneumonia (40%), and radiolucency of the long bones (60%). In some cases there is also hepatitis, myocarditis, or encephalitis which may be fatal (KORONES, RUDOLPH).

The chronicity of infection is generally attributed to the greater susceptibility of embryonic tissue and relative immaturity of immunological responses in the developing fetus. There is no detectable antibody response in the fetus until after mid-pregnancy, when both passive and early active immunoglobulins have been found. The possible impairment of immune defenses should be studied further as a contributing factor in persistent infection with rubella virus. It is conceivable that in the fetus and newborn poorly functioning cellular immune mechanisms or interferon production, which has been shown to be the case in mice (HEINEBERG), might play a major role in the chronicity of rubella infection.

For adults, acute rubella is a mild, one-to-four-day exanthem. Infection usually confers long-lasting immunity with the persistence of high levels of neutralizing antibodies years after infection. Postauricular and/or suboccipital lymphadenopathy is frequently present even when rash does not occur. A second attack is probably a rare occurrence. Among women of childbearing age, approximately one-third to two-thirds of infections occur without rash.

Obviously the importance of rubella as a public health problem is directly proportional to the incidence of congenital rubella. Diagnosis of the latter can be made on the basis of the characteristic abnormalities, a history of maternal rubella, and unusual clinical findings reported in the newborn period. Laboratory data such as isolation of the virus from the placenta, posterior nasopharynx, urine or cerebrospinal fluid of the infant, and positive serologic findings are of great assistance in confirming the diagnosis.

Epidemiology of Rubella

In a survey of 30,000 pregnant women in the Collaborative Perinatal Research Study, it was found that the frequency of clinical rubella was approximately 1:1,000 during non-epidemic years (WHITE, 1966). In the course of the 1964 epidemic, however, the incidence increased to approximately 22:1,000 (Fig. 1). In a recent serological study of 500 pregnant women during the 1964

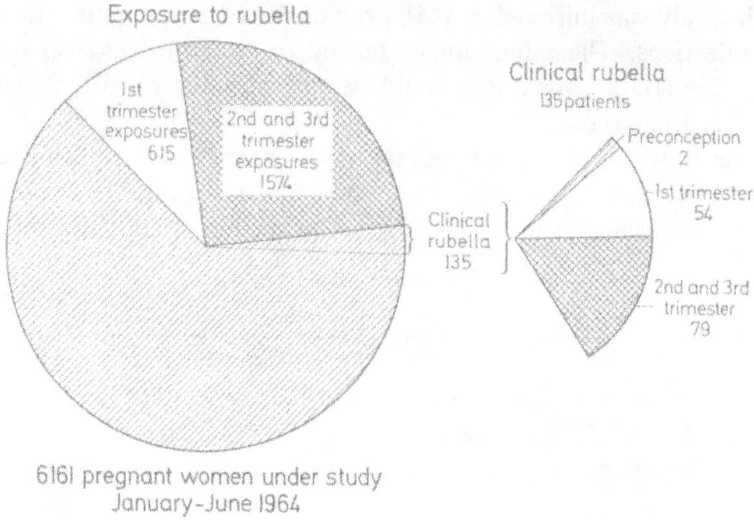

Fig. 1. Incidence of clinical rubella in 6,161 pregnant women. From Perinatal infections affecting the developing fetus and newborn by J. L. SEVER, The prevention of mental retardation through control of infectious diseases (U.S. Department of Health, Education, and Welfare)

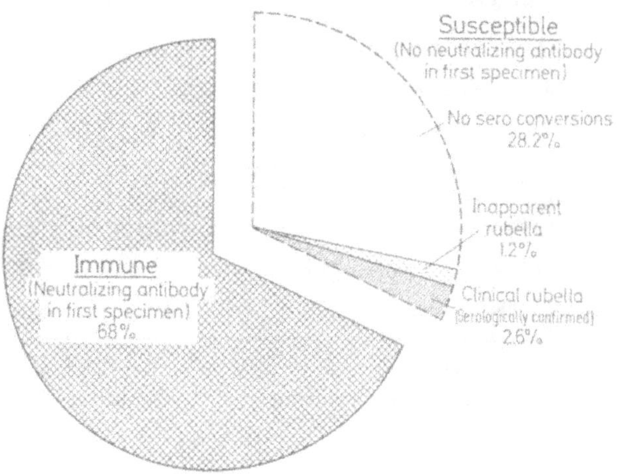

Fig. 2. Results of serological study of 500 pregnant patients during 1964 rubella epidemic. From Perinatal infections affecting the developing fetus and newborn by J. L. SEVER, The prevention of mental retardation through control of infectious diseases (U.S. Department of Health, Education, and Welfare)

epidemic (SEVER, 1966a), 68% were serologically immune, 2.6% had serologically confirmed clinical rubella, and 1.2% had subclinical infection which was detected serologically (Fig. 2).

The frequency of congenital rubella as compared to reported clinical viral infections during pregnancy is summarized in Fig. 3. The analysis is based on a broad survey of approximately 30,000 pregnancies (SEVER, 1968). Seroepidemiological investigations conducted since the Collaborative Perinatal

Research Study was initiated in 1958 provide data for minimum estimates of certain infections. The minimum frequency of confirmed clinical cases per 10,000 in the study population would be: mumps, 10; rubella, 8; varicella-zoster, 5; rubeola, 0.6.

Rubella is worldwide in distribution and it appears that epidemics usually occur at regular intervals ranging from 5 to 10 years. Several large prospective studies are furnishing new data on the epidemiology of this disease and, as

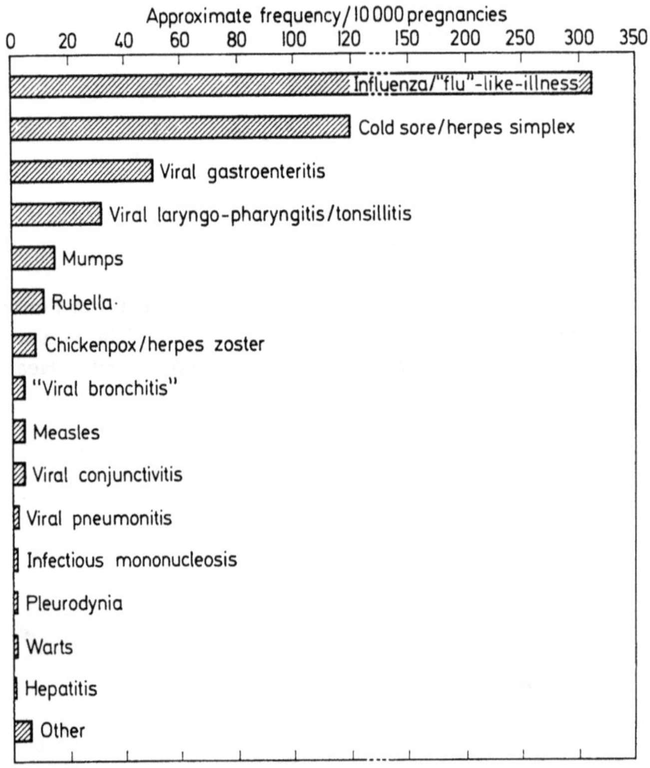

Fig. 3. Frequency of reported clinical viral infections during pregnancy. From Intrauterine viral infections by J. L. SEVER, and L. R. WHITE, Ann. Rev. Med., Chap. 19 (1967)

one would expect, their data indicate that the attack rates depend upon the population being observed, and are affected by numerous factors such as geographic location, season of the year, age groups and interval since the last epidemic.

Immunological Responses

Along with infection, children with congenital rubella also have neutralizing antibody in their sera in utero which persists after birth, apparently for life (WELLER, 1964). This immunological paradox of simultaneous infection and antibody is still unexplained.

The antibody response in acquired rubella differs from that in congenital infection. In the acute (acquired) form of infection, hemagglutination inhibition (HAI), neutralizing (Neut), fluorescent (FA), and complement-fixing (CF) antibodies appear shortly after the onset of rash (SEVER, 1966b). Neut, HAI, and FA antibodies persist indefinitely while CF antibody gradually decreases and tends to disappear completely. With chronic (congenital) rubella, Neut antibody has been detected in the fetus at about the fifth month of gestation and thereafter (BANATVALA). At birth the child has Neut, HAI, FA, and CF titers similar to those of the mother (SEVER, 1966b). During the first few months of life there is a reduction in CF antibody whereas the other antibodies remain detectable and apparently persist for life in most cases.

Antibodies appear to have little if any direct role in the termination of infection (MONIF, ALFORD). The prolonged persistence of congenital infection presumably indicates an impairment of mechanisms concerned with recovery from viral infections, among which special attention should be given to cellular immune mechanisms including delayed hypersensitivity, and production of interferon/intracellular antiviral protein.

Delayed hypersensitivity is an important immune mechanism and although there is no accepted in vitro test to assay delayed-type immunity to rubella virus, lymphocyte transformation has been used to study cellular response in congenital rubella. MONTGOMERY et al. have indicated that peripheral lymphocytes of newborns with the disease show a depressed response to the mitogenic effect of phytohemagglutinin (MONTGOMERY). This depression could be observed only during active infection and was readily duplicated in normal cells exposed to rubella virus prior to the addition of the mitogen. The same phenomenon was noted in other studies, in addition to the demonstration of prolonged impairment of lymphocyte transformation to certain antigens following primary immunization (WHITE, 1968a). While lymphocytes of normal children acquired responsiveness to vaccinia and diphtheria after exposure to these antigens, those of children with congenital rubella failed to do so. These results might lead to a general hypothesis that the immunologic competence of peripheral lymphocytes is altered by congenital rubella being gradually recovered with maturation.

The interferon system is a strong candidate for a possible defective mechanism contributing to the persistent infection. This view is supported by experiments showing the absence of interferon in the serum or urine of patients with congenital rubella syndrome (DESMYTER). It would appear, however, that tissue cultures derived from embryos or newborns infected during intrauterine life persist chronically infected in vitro, but are unable to produce measurable amounts of interferon (RAWLS).

Unfortunately, only limited studies have been done to determine whether or not an impairment of the interferon mechanism exists in congenital rubella and, if so, to what extent this condition would be involved in the maintenance of persistent infection (RAWLS). An animal model system would be an important and necessary tool to elucidate the pathogenesis of chronic neonatal rubella infection. The newborn ferret is known to become readily chronically infected (FABIYI), and this animal is presently being used for studies of circulating interferon and intracellular antiviral protein following rubella virus inoculation.

Since many infants with rubella begin producing antibody to the virus antigen weeks before birth, an increase in serum IgM levels is frequently found. This elevation in the serum IgM concentration in 50% or more of such infants has some immediate diagnostic value when determined in the newborn period but cannot be relied upon to be increased in all cases (SEVER, 1967a).

Since the embryo is synthesizing IgM antibodies, it might be reasonable to expect IgG production prior to birth as well. Moreover, if fetal IgG is transferred to maternal circulation, the mother might react to the foreign protein by producing her own antibody. A possible subsequent perfusion of the embryo with maternal anti-IgG antibodies could result in serious consequences such as immunoglobulin dyscrasia or abnormalities in the development of fetal lymphoid tissue. This possibility was investigated in 24 pregnant women with rubella (WHITE, 1968b). It was found that anti-IgG activity was 3 times as great in the delivery serum of mothers who were infected as in uninfected mothers. However, this increase in frequency was similar to that expected in non-pregnant persons with a recent history of rubella infection, and mercaptoethanol treatment abolished the anti-IgG activity, indicating that the condition was associated with IgM rather than cross-placental fetal IgG.

Serological Methods, Detection and Propagation of the Virus

The obvious importance of serological determinations for rubella in clinical medicine should lead experienced laboratories to employ standard techniques and to repeat all determinations under code.

Rubella antibody determination was introduced in 1962 with the indirect neutralization test (PARKMAN, 1962; WELLER, 1962). Since then, three additional serologic tests for rubella have been developed: FA, CF and HAI. The Neut test is reliable and titers determined by this method correlate well with donors' susceptibility or resistance to infection (SCHIFF, 1965). The FA can be performed in hours, whereas the Neut takes days. However, FA is not as sensitive for determining the development of antibody as the latter. Similarly the CF test is rapid and easy to perform and detects the development of antibody with great accuracy (SEVER, 1967a). Unfortunately, both CF and FA tests show loss of detectable antibody or weakening of fluorescence with specimens taken more than a few months after infection. This factor reduces their value in determining past infection and susceptibility. The HAI is the most recent and perhaps the most accepted test (STEWART). This technique is rapid to perform and consistently gives much higher titers than the other methods.

For serological diagnosis of acute acquired rubella, all four tests are quite suitable. However, due to rapidity of performance, HAI and CF have some advantage over the others. For serological diagnosis of congenital rubella syndrome, HAI, Neut, or FA tests may be used. A significant HAI and CF titer in children with chronic rubella at 7 months to 1 year of age usually indicates a severe infection. To determine individual susceptibility to rubella virus, the HAI and Neut tests are most useful since these antibodies seem to persist for life.

Methods used for the performance of these tests may differ slightly from laboratory to laboratory, but in general, they are essentially the same. Neut tests are based on the enterovirus interference method in primary African green monkey kidney tissue culture (AGMK) (SCHIFF, 1963), or by techniques employing direct cytopathic changes (CPE) in RK-1 (Hull), or in RK-13 (Beale) tissue culture systems. CF tests are performed with antigen prepared by infecting BHK-21 tissue cultures, and utilizing either the micro or the macro technique (SEVER, 1962a). FA determination is easily carried out according to the method described by BROWN and MAASSAB (BROWN) utilizing chronically infected LLC-MK_2 cell line. The HAI test is indeed a simple test once its variables are appropriately controlled. Variations in this test may be caused by modification of the method of treatment of the antigen, the concentration of red blood cells used, the pH of the test, treatment of the sera and employment of macro or micro methods.

Rubella virus can be readily isolated from a variety of specimens from patients with congenital or acquired infection. Most body fluids and tissues were found to reveal the presence of virus; nasopharynx washings or swabs are particularly useful for the detection of the agent.

Basically there are two techniques for rubella virus isolation: the direct technique and the indirect interference method. WELLER and NEVA (WELLER, 1962) first introduced the direct method in 1962 when they described CPE of rubella virus in primary cultures of human amnion cells. Presently the spectrum of tissue culture systems capable of supporting growth of rubella with characteristic CPE, after adaptation, is quite large. These include the RK-13 line (MCCARTHY), GMK-AH-1 line (GUNALP), SIRC line (LEERHOY), BHK-21/WI-2 line (VAHERI), and many others. In primary cells rubella virus very seldom produces CPE (VAHERI), and when CPE is present the changes are slow to appear and difficult to detect.

The indirect technique is based on the interfering action of the virus to superinfection with enteroviruses (usually ECHO-11 and Coxsackie A-9) in AGMK. This blocking action can be abolished with immune rubella serum which is the basis for the Neut test. More recent studies have shown that rubella virus can be propagated and can produce interference to a number of superinfecting viruses in a variety of cell cultures (PARKMAN, 1964).

As the number of tissue culture systems sensitive to the virus broaden, different and more convenient serological techniques are developed. Each system has its advantages and disadvantages, but generally speaking, the interference technique in AGMK has proven to be the most sensitive for recovering virus from crude specimens (SCHIFF, 1966, HORTA-BARBOSA). Comparative data obtained from studies in which fresh field specimens known to be positive for rubella were inoculated simultaneously into AGMK and RK-13 are presented in Table 1.

Table 1. *Comparison of rubella isolations from throat swab specimens: African green interference method and RK-13 direct*

	Interference method	RK-13
No. positive	24	5
No. negative	*18*	*37*
Total	42	42

From Rubella: Recent laboratory and clinical advances by G. M. SCHIFF, and J. L. SEVER: Progress in Prog. Med. Virol. **8**, 30—61 (1966).

A number of attempts were made to produce rubella in laboratory animals (SEVER, 1962a; HEGGIE; PARKMAN, 1965). SEVER et al. (SEVER, 1962b) did not observe clinical symptoms in rhesus monkeys infected with tissue culture grown rubella virus which would produce clinical disease in human volunteers. HEGGIE and ROBBINS (HEGGIE) reported similar results.

More recently, studies on the pathogenesis of the disease in ferrets indicate that infection follows inoculation by different routes, preferably the intranasal (SCHIFF, 1964). Virus multiplies in various organs and is easily recovered from them, which encourages the use of ferrets for vaccine and drug efficacy studies. Another indication that the ferret might become the non-human in vivo model of choice for rubella studies is the prompt development of chronic infection by newborn animals following virus inoculation (FABIYI, 1967). However, an in utero model system to study congenital rubella and its teratogenic mechanism is still not confirmed. Attempts to produce congenital infection in ferrets have been promising. Studies in cynomolgous monkeys, rats and rabbits appear promising but await extension and confirmation.

Immunologically many species respond to rubella virus infection. Usually rabbits, monkeys and ferrets are used to make hyperimmune serum (PARKMAN, 1962; WELLER, 1962; SEVER, 1962a; SCHIFF, 1964; PLOTKIN, 1963).

Control and Prevention

The obvious solution for the rubella problem is the development of an antigen which would prevent the occurrence of epidemics. Intensive efforts have been made in the last few years and hopefully a safe and effective vaccine will be available before the next major epidemic. Both killed and live attenuated preparations have been studied. To date the strongest candidate is the attenuated live HPV-77 strain developed by PARKMAN and MEYER (PARKMAN, 1966; MEYER). The HPV-77 rubella virus (High Passage Virus-77th passage in AGMK) shows several "markers", differing from the low passage virus by producing more interferon, rapid CPE in RK-13, and no viremia or virus in the nasopharynx of monkeys. After extensive safety testing, a vaccine trial with rubella-susceptible females demonstrated good antibody response with no signs of clinical disease, and, more important, none of the contact children became infected. These findings indicate that attenuation has been produced to a point where vaccinees demonstrate pharyngeal shedding of virus, but no transmission to susceptible contacts. However, the susceptibility of the fetus to live rubella virus vaccine remains an unanswered question. Furthermore, more proof is necessary that the virus shed by the vaccinees is not transmissible to large numbers of pregnant susceptibles. Thus far only one preliminary report by KATZ (KATZ) implies non-transmission of HPV-77 vaccine to the fetuses of 2 vaccinated pregnant women.

As a practical weapon against rubella, since the possible hazard of transmission of virus to the fetus has not yet been established, the use of live vaccines could be restricted to children in the prepubertal period if it can be determined that long term immunity is produced. Even then, it is important to remember that the pregnant population would be at risk unless the non-communicability of the virus can be definitely demonstrated. While these questions remain, it is appropriate to think of inactivated vaccines and gamma globulin as supplements to a future safe live vaccine for the control and prevention of rubella.

Besides the HPV-77 strain, other live attenuated strains have been reported. CABASSO et al. (CABASSO) prepared a live vaccine derived from the 50th passage level of the M-33 strain 3 in AGMK. Clinical trials with this preparation gave similar results to those described for HPV-77. Equivalent results were reported by STOKES (STOKES) with the use of a rubella strain serially passed in duck embryo tissue culture. PLOTKIN et al. have described clinical trials using two rubella virus strains (PLOTKIN, 1967), one isolated from the urine of a patient and passaged serially (21 times) in primary rabbit kidney tissue culture (Cendehill-21), the other isolated from a fetus and passaged successively 21 times in WI-38 cell line. These strains produced good antibody responses and contact controls did not seroconvert or develop clinical illness. However, rash and lymphadenopathy were observed in some of the vaccinated children. GITNICK et al. (GITNICK) suggested the use of RIF-free chick embryo tissue culture for the preparation of rubella live vaccine, which is most desirable since this tissue is free of contaminating agents and has been largely administered in live vaccines. Of the tissue reported to date for live vaccine studies only two have been licensed for general use in live parenteral vaccines in the United States: chick embryo and canine kidney. Both are being investigated at the present time as sources of live attenuated rubella vaccine. African green monkey kidney has the well known disadvantage of harboring latent simian viruses and other agents which may be oncogenic.

In summarizing the various approaches for the control and prevention of rubella, four main points should be emphasized: 1) the use of accurate serological techniques in determining individual susceptibility. If a woman is immune to the virus she need not fear exposure during pregnancy. Neut and HAI tests are quite reliable for this purpose and one test may be used to confirm the other. The CF test is of great value in documenting clinical disease in pregnant women exposed to rubella as well as in detecting subclinical infection with the use of appropriately paired sera. 2) Although gamma globulin is not entirely effective in preventing rubella syndrome defects, it has been proven that its use does reduce the incidence of clinical disease (Green, Brody). Gamma globulin antibody titers determined with the HAI and Neut methods (Table 2) indicate the need for intensive studies on the

Table 2. *Rubella antibody. Gamma-globulin — Rubella antibody titers*

Lot	Gamma-globulin		Population		Antibody titer	
	source	concentration (%)	number of donors	location of donors	hemagglutination inhibition	neutralization
1	Com.[a] A	16.5	8,000	Japan	2,048	256
2	Com. A	16.5	8,000	Japan	2,048	512
3	Com. A	16.5	8,000	Japan	4,096	2,048
4	Com. B	16.5	13,000	Urban East U.S.	4,096	1,024
5	Com. C	12.0	[b]	Sweden	2,048	256
6	Com. C	12.0	[b]	Sweden	2,048	512
7	DBS	16.5	Rubeola reference		2,048	512
8	Convalescent	12.0	[b]	Sweden	8,192	4,096
9	Convalescent	16.5	85	Illinois	8,192	4,096

[a] Commercial standard gamma-globulin.
[b] Not available.

From Rubella antibody determinations by J. L. Sever, D. A. Fuccillo, G. L. Gitnick, R. J. Huebner, M. R. Gilkeson, A. Ley, N. Tzan, and R. G. Traub, Pediatrics **40**, 789—797 (1967).

efficacy of this product for the prevention of rubella in pregnant women (Sever, 1967b). 3) An effective killed vaccine assuring long lasting immunity could in itself control the problem of rubella. However, a killed vaccine conferring even a short-term immunity would still be valuable for protecting pregnant women during epidemics of the disease. 4) A safe live attenuated vaccine seems to be the major prophylactic measure now in sight against the rubella syndrome since live antigens produce active long lasting immunity.

Summary

Rubella was shown to produce birth defects a quarter of a century ago. A high incidence of malformations is associated with rubella infection in the first trimester of pregnancy, and severe chronic infection of the newborn has also been demonstrated. Rubella epidemics of worldwide distribution generally occur in intervals of 5 to 10 years.

The role of immune and non-immune mechanisms in the pathogenesis of the disease was discussed, and an immuno-serological analysis of congenital and acquired rubella was attempted.

A variety of reliable laboratory methods is now available for the isolation, propagation and neutralization of the virus. Available serological tests, including FA, HAI, CF, and Neut antibody determinations, are of great importance in the diagnosis, control and prevention of the disease.

Animal studies indicate that subclinical and chronic rubella are readily produced in ferrets which provide a useful animal model system for the study of rubella.

Rapid progress is being made on testing of candidate vaccines. To date, most live vaccines produce good antibody response and very little illness.

References

ALFORD, C. A., F. A. NEVA, and T. H. WELLER: Virologic and serologic studies on human products of conception after maternal rubella. New Engl. J. Med. **271**, 1275—1281 (1964).

BANATVALA, J. E., D. M. HORSTMANN, M. C. PAYNE, and L. GLUCK: Rubella syndrome and thrombocytopenic purpura in newborn infants: clinical and virologic observations. New Engl. J. Med. **273**, 474—478 (1965).

BEALE, J. A., G. C. CHRISTOFINIS, and I. G. FURMINGER: Rabbit cells susceptible to rubella virus. Lancet **1963 II**, 640.

BRODY, J. A., J. L. SEVER, and G. M. SCHIFF: Prevention of rubella by gamma globulin during an epidemic in Barrow, Alaska, in 1964. New Engl. J. Med. **272**, 127—129 (1965).

BROWN, G. C., H. F. MAASSAB, J. A. VERONELLI, and T. J. FRANCIS: Rubella antibodies in human serum: detection by the indirect fluorescent-antibody technique. Science **145**, 943—945 (1964).

CABASSO, V. J., M. R. STEBBINS, S. KARELITZ, C. P. CERINI, J. M. RUEGSEGGER, and M. STILLERMAN: Attenuation of rubella virus: studies in monkeys and man. J. Lab. clin. Med. **70**, 429—441 (1967).

DESMYTER, J., W. E. RAWLS, J. L. MELNICK, M. D. YOW, and F. F. BARRETT: Interferon in congenital rubella: response to live and attenuated measles vaccine. J. Immunol. **99**, 771—777 (1967).

FABIYI, A., G. L. GITNICK, and J. L. SEVER: Chronic rubella virus infection in the ferret (Mustela putorius fero) puppy. Proc. Soc. exp. Biol. (N.Y.) **125**, 766—771 (1967).

GITNICK, G. L., D. A. FUCCILLO, J. L. SEVER, and R. J. HUEBNER: Progress in rubella vaccine development: review and studies of growth of rubella in chick embryo tissue culture. Amer. J. publ. Hlth **58**, 1237—1247 (1968).

GREEN, R. H., M. R. BALSAMA, J. P. GILES, S. KRUGMAN, and G. S. MIRICK: Studies of the natural history and prevention of rubella. Amer. J. Dis. Child. **110**, 348—365 (1965).

GREGG, N.: Congenital cataract following german measles in the mother. Trans. ophthal. Soc. Aust. **3**, 35 (1941).

GUNALP, A.: Growth and cytopathic effect of rubella virus in a line of green monkey kidney cells. Proc. Soc. exp. Biol. (N.Y.) **118**, 85—90 (1965).

HARDY, J. B., G. H. MCCRACKEN, M. R. GILKESON, and J. L. SEVER: Adverse fetal outcome following maternal rubella after the first trimester of pregnancy. J. Amer. med. Ass. (in press).

HEGGIE, A. D., and F. C. ROBBINS: Experimental rubella in rhesus monkeys. Proc. Soc. exp. Biol. (N.Y.) **114**, 750—754 (1963).

HEINEBERG, H., E. GOLD, and F. C. ROBBINS: Differences in interferon contents in tissues of mice of various ages infected with Coxsackie B-1 virus. Proc. Soc. exp. Biol. (N.Y.) **115**, 947—953 (1964).

Horta-Barbosa, L., and J. Warren: The comparative sensitivity of tissue cultures to rubella virus: use of guinea pig cells for virus titration. J. Lab. clin. Med. (in press).

Hull, R. N., and G. Butorac: The utility of rabbit kidney cell strain, LLC-RK to rubella virus studies. Amer. J. Epid. **83**, 3, 509—517 (1966).

Katz, S.: Presentation. Amer. Ped. Soc. Meeting, Atlantic City, N. J., 1967.

Korones, S. B., L. E. Ainger, G. R. Monif, J. Roane, J. L. Sever, and F. Fuste: Congenital rubella syndrome: new clinical aspects with recovery of virus from affected infants. J. Pediat. **67**, 166—181 (1965).

Leerhoy, J.: Cytopathic effect of rubella virus in a rabbit cornea cell line. Science **149**, 633 (1965).

Lundstrom, R.: Rubella during pregnancy: a follow-up study of children born after an epidemic of rubella in Sweden, 1951, with additional investigations on prophylaxis and treatment of maternal rubella. Acta Paediat. (Uppsala) **51**, Suppl. 133, 1—110 (1962).

Maton, W. G.: Some account of a rash liable to be mistaken for scarlatina. Medical Tr. Roy. Coll. Phys., London **5**, 149 (1815).

McCarthy, K., C. H. Taylor-Robinson, and S. E. Pillinger: Isolation of rubella virus from cases in Britain. Lancet **1963 II**, 593—598.

Meyer, H. M., Jr., P. D. Parkman, and T. C. Panos: Attenuated rubella virus: II. Production of an experimental live-virus vaccine and clinical trial. New Engl. J. Med. **275**, 575—580 (1966).

Michaels, R. H., and G. W. Mellin: Prospective experience with maternal rubella and the associated congenital malformations. Pediatrics **26**, 200—209 (1960).

Monif, G. R., J. H. Hardy, and J. L. Sever: The lack of association between the appearance of complement fixing antibodies and the recovery of virus in a child with congenital rubella. Pediatrics **39**, 289—291 (1967).

Montgomery, J. R., M. A. South, W. E. Rawls, J. L. Melnick, G. B. Olson, P. B. Dent, and R. A. Good: Viral inhibition of lymphocyte response to phytohemagglutinin. Science **157**, 1068—1070 (1967).

Parkman, P. D., E. L. Buescher, and M. S. Artenstein: Recovery of rubella virus from Army recruits. Proc. Soc. exp. Biol. (N.Y.) **111**, 225—230 (1962).

— — — J. M. McCown, F. K. Mundon, and A. D. Druzd: Studies of rubella. I. Properties of the virus. J. Immunol. **93**, 595—607 (1964).

— H. M. Meyer Jr., R. L. Kirschstein, and H. E. Hopps: Presentation. Amer. Ped. Soc. meeting, Atlantic City, N. J., 1966.

— P. E. Phillips, and H. M. Meyer Jr.: Experimental rubella virus infection in pregnant monkeys. Amer. J. Dis. Child., **110**, 390—394 (1965).

Plotkin, S. A., J. A. Dudgeon, and A. M. Ramsey: Lab. studies on rubella and the rubella syndrome. Brit. med. J. **1963 II**, 1296—1299.

— J. D. Farquhar, S. Katz, and T. H. Ingalls: Discussion of rubella vaccines. Pan Amer. Hlth Org., Sci. Publ. **147**, 405—408 (1967).

Rawls, W. E., and J. L. Melnick: Rubella virus carrier cultures derived from congenitally infected infants. J. exp. Med. **123**, 795—816 (1966).

Rudolph, A. J., M. D. Yow, A. Phillips, M. M. Desmond, R. J. Blattner, and J. L. Melnick: Transplacental rubella infection in newly born infants. J. Amer. med. Ass. **191**, 843—845 (1965).

Schiff, G. M., J. L. Sever, and R. J. Huebner: Experimental rubella: clinical and laboratory findings. Arch. intern. Med. **116**, 537—543 (1965).

— — — Rubella Virus: neutralizing antibody in commercial gamma globulin. Science **142**, 3588, 58—60 (1963).

— — — Experimental rubella infection in ferrets. Clin. Res. **12**, 245 (1964).

— — Rubella: recent laboratory and clinical advances. Prog. med. Virol. **8**, 30—61 (1966).

SEVER, J. L.: Application of a microtechnique to viral serological investigations. J. Immunol. **88**, 320—329 (1962b).

—, and H. W. BERENDES: Cord blood gamma M as a screening test for congenital viral infections. Pediat. Res. **1**, 217 (1967a). Abstract.

— D. A. FUCCILLO, G. L. GITNICK, R. J. HUEBNER, M. R. GILKESON, A. LEY, N. TZAN, and R. G. TRAUB: Rubella antibody determinations. Pediatrics **40**, 789—797 (1967b).

— R. J. HUEBNER, A. FABIYI, G. R. MONIF, G. CASTELLANO, C. L. CUSUMANO, R. G. TRAUB, A. LEY, M. R. GILKESON, and J. M. ROBERTS: Antibody responses in acute and chronic rubella. Proc. Soc. exp. Biol. (N.Y.) **122**, 513—516 (1966b).

— G. R. G. MONIF, C. L. CUSUMANO, G. M. SCHIFF, and R. J. HUEBNER: Clinical and subclinical rubella following intradermal inoculation of rubella virus into human volunteers. Amer. J. Epid. **84**, 163—166 (1966a).

— G. M. SCHIFF, and R. G. TRAUB: Rubella virus. J. Amer. med. Ass. **182**, 663—671 (1962a).

—, and WHITE, L. R.: Intrauterine viral infections. Ann. Rev. Med. **19**, 471—486 (1968).

STEWART, G. L., P. D. PARKMAN, H. E. HOPPS, R. D. DOUGLAS, J. P. HAMILTON, and H. M. MEYER JR.: Rubella-virus hemagglutination inhibition test. New Engl. J. Med. **276**, 554—557 (1967).

STOKES, J., JR., R. E. WEIBEL, E. D. BUYNAK, and M. R. HILLEMAN: Clinical and laboratory tests of Merck strain live attenuated rubella virus vaccine. Pan Amer. Hlth Org., Sci. Publ., **147**, 402—405 (1967).

VAHERI, A., W. D. SEDWICK, and S. A. PLOTKIN: Growth of rubella virus in BHK-21 cells. I. Production, assay, and adaptation of virus. Proc. Soc. exp. Biol. (N.Y.) **125**, 1086—1092 (1967).

WELLER, T. H., C. A. ALFORD, and F. A. NEVA: Retrospective diagnosis by serological means of congenitally acquired rubella infection. New Engl. J. Med. **270**, 1039—1041 (1964).

—, and F. A. NEVA: Propagation in tissue culture of cytopathic agents from patients with rubella-like illness. Proc. Soc. exp. Biol. (N.Y.) **111**, 215—225 (1962).

WHITE, L. R., S. LEIKEN, O. VILLAVICENCIO, W. ABERNATHY, G. AVERY, and J. L. SEVER: Immune competence in congenital rubella: Lymphocyte transformation, delayed hypersensitivity, and response to vaccination. J. Pediat. **73**, 229—234 (1968a).

— F. W. ROSA, and J. L. SEVER: Study on cerebral palsy, mental retardation and other neurological and sensory disorders of infancy and childhood. Abstracts of Presentation, p. 25, Washington, D. C., 1966.

— J. L. SEVER, and F. P. ALEPA: Maternal and congenital rubella before 1964: Frequency, clinical features, and search for isoimmune phenomena. Pediatrics (in press) (1968b).

Departments of Microbiology and Pediatrics, State University of New York at Buffalo, School of Medicine, and Laboratories of Bacteriology, Children's Hospital and Roswell Park Memorial Institute, Buffalo, New York

Endotoxins and the Immune Response*

ERWIN NETER

Table of Contents

Introduction

This review deals with immunologic aspects of the effects of endotoxins in susceptible hosts. The subject matter is divided into three sections as follows: (1) biologic effects of endotoxins and immunologic implications, (2) endotoxins as adjuvants, and (3) endotoxins as immuno-suppressants. Not considered is the extensive literature on the antigenic make-up of endotoxins, characterizing the numerous O and R antigens, and endotoxins as immunogens.

Biologic Effects of Endotoxins and Immunologic Implications

Endotoxins, products of gram-negative bacteria, are macromolecules composed of polysaccharide, lipid, and protein or polypeptide. They are integral parts of the outer triple-layered envelope, which, in turn, is a portion of the cell wall of gram-negative bacteria (BLADEN et al., 1967; MERGENHAGEN et al., 1966). When isolated from smooth strains, they characterize the O antigenic specificity of species, sero-groups, or serotypes of these microorganisms, and it is the polysaccharide moiety that carries the antigenic determinants. Significant progress has been made during the past few years in the identification of the various haptenic determinants, largely due to the investigations of WESTPHAL, LÜDERITZ, STAUB, and KAUFFMANN (cf. LÜDERITZ et al., 1966; LÜDERITZ et al., 1968; and ØRSKOV et al., 1967). Endotoxins are produced also by rough strains, and again it is the polysaccharide that is responsible for the serologic R specificity.

In susceptible animals, these macromolecules produce an extraordinary array of physiologic and pathologic effects. Equally remarkable is the host range, from invertebrates (LEVIN, 1967) to primates (GILBERT, 1962), and

* The author's research was supported by Grant AI00658, National Institute of Allergy and Infectious Diseases, U.S.P.H.S.

including man as well. In human subjects the hemorrhagic lesions of the Waterhouse-Friderichsen syndrome as well as shock associated with systemic infection due to endotoxin-producing bacteria closely resemble similar phenomena seen in experimental animals. It must be emphasized, however, that marked differences exist between various animal species with regard to their susceptibility to endotoxin and to the lesions it engenders. In one such study by BERCZI et al. (1966) a lethal dose of *E. coli* endotoxin (mg/kg) was 0.025 for calf, 3 for rabbit, and >50 for chicken. In another study (KIM and WATSON, 1966), the LD_{50} of intravenously injected endotoxin was approximately 500 μg/kg for the rabbit, 100 μg/kg for the newborn piglet, and 10,000 μg/kg for adult Balb/Sy mice 10 weeks old. The complexity of the problem is well illustrated by the observations of PIERONI and LEVINE (1968) who showed that the CFI mouse strain, which is at least as susceptible as the CFW strain to the lethal effect of typhoid endotoxin, is nevertheless refractory to its histamine sensitizing effect. It is of interest to point out, too, that the route of administration is significant, the intravenous route being more effective than the intraperitoneal route in the rabbit but not in the mouse. Of equal interest is the observation that raising the environmental temperature of experimental animals from 22—23°C to 37°C results in a striking decrease in the LD_{50} dose: in the mouse from 240 μg to 4 (KASS et al., 1964). As will be discussed below, certain of the toxic effects of endotoxins have immunologic features, involving antibodies and/or complement, and so does tolerance induced by repeated administration of these macromolecules.

The physiologic and pathologic effects produced by endotoxins include the following, often inter-related, reactions, summarized here, for the sake of convenience, into ten major groups:

(1) Fever; endotoxins are highly effective pyrogens. The minimal effective dose depends upon the particular endotoxin and the criteria used for documentation of the febrile response. HASKINS et al. (1961) produced fever in the rabbit with as little as 0.00015 μg/kg. Human subjects, too, are susceptible, a total dose of 0.05 to 0.1 μg producing a temperature elevation in man (WESTPHAL, 1957). In the pathogenesis of fever, the release of endogenous pyrogen plays a significant role (cf. COLLINS and WOOD, 1959; ATKINS and SNELL, 1965). That endogenous pyrogen is not the sole mediator of fever is evident from the facts that cerebral injection of endotoxin is more effective than intravenous administration and that this febrile response does not depend upon the presence of circulating granulocytes (DU BUY, 1966).

(2) The localized and generalized Shwartzman and Sanarelli reactions (DACHY, 1967), increased reactivity to epinephrine resulting in hemorrhagic lesions (NETER et al., 1960), and to serotonin (WEINER and ZWEIFACH, 1966), hemorrhagic tumor necrosis (MIHICH et al., 1961), and placental injury with fetal wastage or pre-mature delivery (McKAY and WONG, 1963), all are produced by endotoxin.

(3) Alterations of the hematopoietic system, such as leucopenia and leucocytosis, and changes in the coagulation and fibrinolytic mechanisms (GANS

and KRIVIT, 1961; TRENTIN, 1967) are effected by these macromolecules. The decrease in the number of circulating lymphocytes and granulocytes is followed by leucocytosis, almost exclusively due to an increase in granulocytes. This secondary granulocytosis depends upon the availability of cells from the bone marrow pool. It is for this reason that endotoxin has been used clinically to gauge marrow granulocyte reserves (HERION et al., 1965). Recent studies by CHERVENICK et al., (1967) indicate that endotoxin in mice causes release of neutrophils from the bone marrow, and the investigations by BOGGS et al. (1968) suggest that in human subjects endotoxin causes the release of a factor in plasma capable of inducing significant neutrophilia. Thrombocytopenia may follow intravenous injection of endotoxin and is not rarely encountered in patients with infection due to endotoxin-producing microorganisms (SPIELVOGEL, 1967). Endotoxin also alters erythropoiesis (FRUHMAN, 1966). Of particular importance are the results of recent investigations on the effects of endotoxin on complement, and the implications of these findings regarding the mode of action of endotoxins are discussed below. It is noteworthy, also, that in susceptible hosts endotoxin may lead to the generation or increased production of C-reactive protein (PATTERSON et al., 1968; PATTERSON and HIGGINBOTHAM, 1965) and of interferon or interferon-like viral inhibitors (OH, 1966).

(4) Changes in metabolism and biochemical homeostasis, such as hypoglycemia, represent yet another group of host reactions. During the past decade BERRY and his associates (BERRY et al., 1959; BERRY et al., 1967) have undertaken a systematic study of metabolic changes engendered by endotoxin, and they have documented that endotoxin in the mouse causes a lowering of the activity of the hormonally inducible liver enzyme tryptophan pyrrolase and, contrariwise, an increase in the activity of a similarly inducible liver enzyme, tyrosine-alpha-ketoglutarate transaminase. These effects must be mediated indirectly, for endotoxin added *in vitro* to liver cells fails to cause a parallel alteration of enzyme activity. Of particular importance, too, is the labilizing effect of endotoxin on lysosomes with attendant alteration of intracellular enzyme levels (WEISSMANN, 1964).

(5) Decrease and/or increase in activity of the RES are among the characteristic endotoxin effects (BENACERRAF and SEBESTYEN, 1957; ARREDONDO and KAMPSCHMIDT, 1963; MILER, 1963; FORBES, 1965).

(6) Changes in nonspecific resistance to various pathogens can be induced by these microbial products. Increased resistance pertains to a variety of infections, including those caused by bacteria (ROWLEY, 1956; DUBOS and SCHAEDLER, 1956; LANDY, 1956), viruses (WAGNER et al., 1959), fungi (KIMBALL et al., 1968), and protozoa (SINGER et al., 1964). It is of interest to note that increased resistance can be produced during the early period of ontogenesis (PETERSON and PAVEL, 1967). Possible mechanisms have been reviewed by WHITBY et al. (1961). Stimulation of resistance has been documented also by lipid derived from endotoxin (MERGENHAGEN et al., 1963). The stimulation of interferon production by endotoxin has been explored as one mechanism

of increased resistance by, among others, YOUNGNER and STINEBRING (1966) and STINEBRING and YOUNGNER (1964).

(7) Endotoxin is an immunogen in suitable animals.

(8) Increase or decrease in the immune response to certain non-self antigens, and alteration of hypersensitivity reactions, are effected by endotoxin.

(9) Circulatory changes, including shock, and finally,

(10) Death may result from endotoxin administration.

In view of the fact that microorganisms producing endotoxin are widespread in nature and that the endotoxin is heat-stable, the possibility of accidental and unintentional contamination of a variety of materials has to be given careful consideration, to avoid erroneous conclusions in experimental studies. As early as 1923, Florence SEIBERT (1923) and SEIBERT and MENDEL (1923) provided strongly suggestive evidence of such contamination of various proteins and even of distilled water. More recently, KESSEL et al. (1966c) detected such contamination in a coliphage preparation and suggest that even some of the commercially available gamma globulin preparations may contain endotoxin.

Many of the phenomena of endotoxicity have indirect mechanisms. It is for this reason that changes observed in the intact animal cannot always be reproduced by exposure of isolated organs or cell populations to endotoxin. Among the organs and tissues studied are cardiopulmonary preparations, isolated intestine, and lungs of guinea pigs (GÄRTNER et al., 1964), isolated liver of the rat (NOLAN and O'CONNELL, 1965), and aortic strip of the rabbit (WEINER and ZWEIFACH, 1966). Reference has been made above to the observation that the liver enzyme changes, studied by BERRY et al. (1959, 1967) are not observed upon addition of endotoxin to liver cells; nor is lysosome labilization effected by *in vitro* addition of endotoxin to isolated lysosomes (WEISSMANN, 1964). Endotoxin administered to mice augments, after initial depression, the phagocytic and bactericidal activity of peritoneal leucocytes to a significantly greater extent than does lipopolysaccharide treatment *in vitro* (WIENER et al., 1965). Certain other alterations, however, are observed *in vitro* upon exposure of cell populations to endotoxin; for example, endotoxin, even at low concentrations (0.005 to 0.05 μg/ml), inhibits the migration of splenic macrophages of guinea pig, rabbit, and man and causes severe damage to large wandering cells (HEILMAN and BAST, 1964, 1967; HEILMAN, 1964, 1968). Mention should be made also of the fact that the lipopolysaccharides readily become attached to erythrocytes and to other cells as well, thus endowing these cells with a newly acquired serologic specificity (SPRINGER and HORTON, 1964; NETER et al., 1956; NETER, 1956, 1965). As a result, antibodies directed against the polysaccharide component of the lipopolysaccharides produce hemolysis *in vitro* and *in vivo* and, conceivably, other types of cell injury as well (SHUMWAY et al., 1963). Thus, indirect immunologic injury can be mediated by these lipopolysaccharide antigens.

It has been clearly established that the protein or polypeptide of the macromolecule is not required for endotoxicity, for lipopolysaccharide also produces the alterations. It appears likely that the lipid A component[1], as originally postulated by WESTPHAL and his associates (1958), plays a role in endotoxicity (e.g. NETER et al., 1960; MIHICH et al., 1961; ARGENTON et al., 1961). Major progress has been made in our understanding of the critically important components of the endotoxin molecule by the study of mutants. KIM and WATSON (1967), utilizing mutants from *Salmonella*, observed that endotoxin, containing mostly lipid A and ketodeoxyoctonate (KDO) and prepared from a mutant deficient in both O- and R-antigens and the backbone sugar, heptose, is biologically active, and they concluded that it is the lipid portion of the molecule, which includes also KDO, that is responsible for endotoxicity. Similarly, the O-polysaccharide is not responsible for complement fixation in guinea pig serum by LPS which leads to the morphological defect or "lesion" on the lipopolysaccharide particle (MERGENHAGEN et al., 1968). Similar conclusions were reached by KASAI and NOWOTNY (1967), who studied the biologic activities of a glycolipid from a heptoseless mutant of *Salmonella minnesota*. The authors point out that this product is as active as the lipopolysaccharide regarding pyrogenicity, Shwartzman reaction, and lethality in chick embryos, but, it is less effective when measured by lethality tests in mice. Further studies, however, will be needed, for KESSEL et al. (1966) observed that endotoxin isolated from a mutant possessing only the backbone was less toxic than endotoxin from smooth and rough strains. Whether it is the isolation procedure, the particular strain, or other conditions that are the basis for these divergent results remains to be determined in the future.

Endotoxin was visualized as a membranous structure by electron microscopy (BLADEN et al., 1967; and SHANDS et al., 1967). ROTHFIELD et al. (1966) suggested a "leaflet structure" for lipopolysaccharide. Electron microscopic studies by SHANDS et al. (1967) suggest that the major morphological determinant of the lipopolysaccharide is the lipid moiety and that the "core" polysaccharide contributes to the trilaminar structure, in contrast to the O-polysaccharide side-chains. RUDBACH, MILNER, and RIBI (1967) have postulated that a minimal molecular size and a critical tertiary or quaternary structury are necessary for endotoxins to elicit the characteristic host reactions; the endotoxin is visualized as a micellar bundle of subunit chains held together by secondary or tertiary forces. It will be of considerable interest to elucidate, by means of electronmicroscopy, the structure of endotoxoids, to be discussed below. It remains for future investigations, also, to identify the components of lipid A that may account for biologic effects of lipopolysaccharide and to learn whether the two concepts, one, assuming a toxic part of the molecule, and the other, emphasizing the critical role of the configuration of the whole macromolecule, rather than being mutually exclusive, may, in fact, supplement each other. It is conceivable that a portion of the endotoxin molecule, perhaps lipid A, is responsible for certain toxic effects, but that the macro-

[1] The term "lipid A" is used in this communication without inferring that it represents a well defined entity. Rather, the term refers to the firmly bound lipids of the endotoxin molecule.

molecule containing this toxic component is required for complete manifestations of biologic activity. This concept would be analagous to that of antibodies, a portion of the immunoglobulin being responsible for combination with antigenic determinants and the remainder for certain related biologic activities, such as complement fixation.

If there is a primary event, on a molecular basis, common to many, if not all, of these physiologic and pathologic changes engendered by endotoxins, such a mechanism has not yet been identified. In view of the multiplicity of host reactions and the differences in susceptibility of various animal species, such an analysis presents obvious difficulties. Based on presently available information, endotoxicity apparently has at least two significant components: (1) certain primary toxic reactions, being probably associated with the seemingly nonantigenic lipid A and (2) reactions of an immunologic type, in the broadest sense of the term, between the antigenic polysaccharide-lipid-protein (or polypeptide) molecule and the host. Support for the concept of a direct pharmacologic effect of endotoxin in susceptible animals, independent of the presence of natural antibodies, has been provided by the elegant studies of KIM and WATSON (1966). These investigators found that germ-free, neonatal, and colostrum-deprived piglets do not possess natural antibodies, detectable by modern and sensitive techniques, and yet these animals are susceptible to endotoxin. In this connection it is of interest to mention that endotoxin in pig serum fixes the same relative percentage of C′ as it does in adult guinea pig serum (SNYDERMAN et al., 1968b). It should be mentioned, however, that WEBB and MUSCHEL (1968) recently described the presence of bactericidal antibodies in the serum of newborn piglets before ingestion of colostrum, although the titers were very low. Sow colostrum was shown to protect these piglets against the lethal effects of endotoxin. Although KIM and WATSON (1966) have observed that the protective antibody resides in the 19S fraction, it is not the O antibody that is effective in passive protection. The specificity of the antibody has not yet been identified, i.e., the antigenic determinant of endotoxin molecules remains to be determined. It would be of interest, also, to learn whether colostrum protects against lipid A, provided this material is lethal to these piglets. Thus, it appears that certain toxic effects produced by endotoxins are not dependent upon the presence of circulating antibodies and, in fact, can be prevented by a yet unidentified antibody. The second mechanism of endotoxicity may be related to the immunologic reactivity of the host. For example, MERGENHAGEN and JENSEN (1962) showed that mice, harboring few, if any, veillonellae, are made resistant to the lethal effects of endotoxin derived from these microorganisms by prior immunization. Several students of endotoxins have long invoked an immunologic basis for many, if not all, manifestations of endotoxicity. STETSON (1964), for the past 15 years, has suggested that hypersentitivity plays a role in endotoxicity, by pointing out that the major effects produced by endotoxin, namely, fever, shock, death, as well as the local and generalized Shwartzman reactions, can all be produced by antigen-antibody interactions in defined systems, and he postulates that "antibodies to endotoxin" play a significant role in these phenomena. Since endotoxin fixes complement components, including components 5 through 9 (ADKINSON et al., 1968; GEWURZ, SHIN, and MERGENHAGEN, 1968), certain host reactions may have a mechanism similar to that of the Arthus reaction. It would be a singularly important advance to have available the identification of the responsible antigenic determinant in order to study decisively the role of these antibodies. KIM and WATSON (1964) reported that papain inactivates the pyrogenic and lethal activities of endotoxin, and they suggest that, if a part of endotoxicity is related to hypersentitivity of the host, a peptide portion of the macromolecule would be the most likely determinant. Additional studies clearly are needed. Such an immunologic pathogenesis may even apply to toxic effects demonstrated with certain cell populations *in vitro*. KESSEL and BRAUN detected cytotoxicity of lipopolysaccharide on guinea pig macrophages that may be mediated by cytophilic antibodies (KESSEL and BRAUN, 1965; KESSEL et al., 1966a). Studies by FREEDMAN et al. (1967b) suggest the participation of the protein part of the whole endotoxin molecule in certain host reactions, the immunologic basis possibly being akin to delayed hypersensitivity. Among the endotoxin manifestations that may have an immunologic basis is thrombocytopenia, possibly related to immune adherence (SPIELVOGEL, 1967). It is noteworthy to mention the recent studies of PAVLOVSKIS and SPRINGER (1967), who have identified an endotoxin-receptor on the membrane of erythrocytes, with which these macromolecules interact. In unpublished experiments carried out in the laboratories of SPRINGER and NETER, it was observed that this lipopolysaccharide-receptor has features of specificity, in as much as it reacts with polysaccharide and another antigen of gram-negative bacteria (common antigen) but not, even in considerably larger amounts, with a common antigen of gram-positive bacteria. As yet, it is not known whether similar receptors are present in other endotoxin-susceptible cells, and what role, if any, these receptor molecules play in the initial stages of endotoxin injury. If, indeed, part of endotoxicity is related to immune mechanisms, secondary effects may be due to the release, following cell injury, of cellular materials, such as oligodeoxyribonucleotides, which, in turn, may profoundly affect host reactions (KESSEL et al., 1966a). The subject of the biodynamic effects of oligonucleotides was admirably discussed by BRAUN and FIRSHEIN (1967). If these considerations of a dual pathogenesis of endotoxicity are valid, then, all studies designed to discern and differentiate these two basic mechanisms should be carried out in parallel with lipid A, endotoxoid, and the total endotoxin complex, clearly a formidable, albeit necessary, undertaking.

Based on the results of studies on the interaction of endotoxin and the complement system, MERGENHAGEN and his co-workers suggest that at least one major effect of endotoxin in producing biologic reactions in susceptible hosts is related to the activation of the complement system without obvious participation of antibodies. As early as 1961, SPINK and VICK (1961) suggested that complement may play a role in endotoxic shock. GILBERT and BRAUDE (1962) observed that lethal doses of endotoxin causes a fall in complement titers in rabbits, in addition to lowering of antibody titers. In a detailed study BLADEN et al. (1967) made the important observation that lesions, similar to these seen on erythrocytes during immune hemolysis, develop both on the surface of *Veillonella alcalescens* as well as on the isolated lipopolysaccharide. These lesions are due to activation of the complement system. The authors suggest that lipopolysaccharide is a substrate for the complement enzymes and that the latter may account for certain biologic effects of endotoxins. GEWURZ et al. (1968), then, documented the fact that lipopolysaccharides isolated from other microorganisms as well fix complement present in normal guinea pig, rabbit, mouse, and human serum. Detoxified endotoxin (endotoxoid) was markedly less active. It is primarily C'3 t that is fixed under these conditions. MERGENHAGEN et al. (1968) also observed that the endotoxin isolated from *S. minnesota* mutant, deficient in heptose and O-polysaccharide, was as active as that of the wild-type strain. As expected, electron microscopic examination of both endotoxins, after treatment with fresh guinea pig serum, revealed numerous characteristic lesions, approximately 90 A in diameter. As reported in the spring of 1968, ADKINSON et al. (1968) made the additional observation that endotoxin, during or after consumption of C'3 and later acting complement components, results in the generation of anaphylatoxin. It remains to be seen whether or not antibody is involved in this phenomenon. In addition to the anaphylatoxin, a chemotactic substance for polymorphonuclear leucocytes is generated (SNYDERMAN et al., 1968a, b). These results from MERGENHAGEN'S laboratory may lead to a more definitive understanding of certain biologic activities of endotoxin.

Attention may be called to the fact that, although an enormous amount of effort has been directed toward the elucidation of endotoxicity, surprisingly limited information is available with regard to the role of endotoxin in infections caused by gram-negative microorganisms. Among these inadequately explored areas are (1) endotoxic activity of intact bacteria in blood and tissues, (2) the release of endotoxin from microbial cells under *in vivo* conditions, and (3) the fate of biologically active material in a variety of infections due to endotoxin-producing pathogens of both animals and man.

The nature of the refractoriness of animals to endotoxin after repeated injections of this material has not been unequivocally identified. It has been known for many years that repeated injections of one endotoxin renders the animal less susceptible to certain toxic effects of the identical (e.g., from *Neisseria*) or seemingly unrelated endotoxin (e.g., from *Salmonella*). BERGER and FUKUI (1963) reported that repeated injections of endotoxin, which result in tolerance, do not preclude the development of increased resistance to experimental infection in mice, assuming that two moieties exist in endotoxin or that two modes of action are involved. Some students of endotoxicity relate this "tolerance" to nonspecific mechanisms, notably altered RES function, but others have provided suggestive evidence of an immune mechanism. GREISMAN et al., (1964) have shown that tolerance of healthy volunteers following daily intravenous injections of bacterial endotoxin was not associated with increased nonspecific activity of RES, in removing labeled aggregated human serum albumin, and they conclude that tolerance in man involves the participation of antibodies resulting in more effective clearance by the RES and/or inactivation of the endotoxin molecule. More recent investigations have revealed that the induction of tolerance by one endotoxin against that from a seemingly unrelated microorganism may yet be based on the presence of antibodies against shared antigenic components. A common antigen between *S. typhosa* and *Pseudomonas* has been postulated to explain cross-tolerance (GREISMAN et al., 1964). KIM and WATSON (1966) and WATSON and KIM (1963) have shown that endotoxin from *Chromobacterium violaceum* induces refractoriness against that of *E. coli* O8 but not in the reversed order. It is not the O antibodies, however, that induce tolerance (KIM and WATSON, 1965). It should be kept in mind that microorganisms share antigenic determinants not only among themselves, an area that has not received adequate and systematic investigation, but bacteria also have antigens in common with higher animals, for example, blood group specificity, studied in great detail by SPRINGER and associates (1966). That endotoxins from *S. typhosa* and *E. coli* O127 differ when one is used for induction of tolerance and the other for challenge was demonstrated by FREEDMAN and SULTZER (1964), and these authors also suggest that antibodies play a role in the development of tolerance, contributing to the removal and/or degradation of the endotoxin. The complexity of the problem is well illustrated by the results of recent studies indicating that there exist at least two distinct mechanisms leading to tolerance: Rabbits treated with endotoxin from three different microorganisms were equally tolerant 24 hours later to one of them; in contrast, after 7 days the rabbits were tolerant only to the endotoxin used for the induction of tolerance. Sera from animals injected 7 days previously transferred tolerance passively, and this was ascribed to the presence of specific antibodies (MILNER and RUDBACH, 1968). On the basis of their observations, WATSON and KIM (1964) and KIM and WATSON (1966) postulate that tolerance is, in fact, the result of immunity to endotoxin. At the present time, however, information is lacking as to the antigenic determinant(s) common to several endotoxins and the particular antibodies, humoral or cell-attached, that are effecting immunity (tolerance). Investigation of the nature or tolerance in various animal species is made even more complex by the fact that significantly different amounts of endotoxin produce similar host manifestations. For example, as mentioned by GREISMAN et al. (1964), the respective doses of a single endotoxin for man and rabbits are 0.007 μg/kg for the former and 0.7 μg/kg for the latter. Attempts to study the role of antibodies in endotoxin tolerance by WOLFF et al. (1965) have revealed that 6-mercaptopurine treatment of rabbits does not interfere with the induction of tolerance and that plasma from these treated animals, presumably poor in antibody content, was as effective in passively transferring tolerance as that from presumably antibody-rich control animals.

The role of the immune status of the host in influencing susceptibility to endotoxin has presented difficulties in evaluation. It is not questioned that antibody directed against some antigenic determinant of the endotoxin molecule can alter *in vivo* responses. Indeed, careful studies by Nowotny et al. (1965) as well as Radvany, Neale, and Nowotny (1966) have clearly shown that mixing of endotoxin and O antibodies prior to injection alters some of the biologic reactions of the host. Rather, it is the documentation that these antibodies are a *conditio sine qua non* for susceptibility to, or resistance against, endotoxicity in various animal species, the identity of the antigen-antibody systems, and the analysis of all pertinent host reactions affected by these antibodies, that are needed for a definitive assessment of the role in endotoxicity of various immunologic events. The difficulties are compounded by the possibility that delayed type hypersensitivity with attendant antibody function of cells may play a role in host susceptibility to endotoxin; it is conceivable, too, that after injection of endotoxin hitherto non-operative antigenic determinants may emerge by *in vivo* alteration of the macromolecule; and, finally, methodology becomes important, for it is clear that all antigen-antibody reactions cannot be documented by most serologic procedures, for example, the common antigen of enteric bacteria (see below) is not precipitated by the corresponding antibody, nor does the latter cause agglutination of bacterial cells that have the antigen on the surface.

At this point, it may be opportune to summarize briefly the concept, advanced by Watson and Kim (1963, 1964), on the immunologic aspects of endotoxicity and tolerance. Toxicity has two components, primary toxicity (part of the fever response, lethality, and tissue damage) is non-immunogenic in nature and secondary toxicity (skin hyperreactivity, Shwartzman reaction, and part of the fever response) is due to hypersensitivity of the host. Tolerance is a form of immunity, based on action of antibody in conjunction with RES.

The term "endotoxin" is a misnomer for two reasons, firstly, because these macromolecules are part of the cell wall rather than of the cytoplasm and, secondly, because they produce, among other reactions, increased resistance to infection that hardly can be termed toxic. Obviously, too, they differ from the classic bacterial toxins, such as diphtheria, tetanus, and botulism toxins, in several aspects. The classic toxins are proteins and the corresponding antibodies abolish toxicity (active and passive immunity). The endotoxins, on the other hand, are lipopolysaccharide in nature, the lipid moiety by itself apparently not being immunogenic and antibodies with O specificity against the polysaccharide part not abolishing toxicity. The term "toxoid" as applied to the derivatives of classic toxins and of endotoxins also has different connotations. With regard to the former, toxoids are nontoxic immunogens. So far as endotoxoids are concerned, at least on an operational basis, the term applies to those endotoxin derivatives which produce diminished toxic "side-effects" but retain their ability to stimulate the desirable nonspecific resistance to infection, enhancement of the immune response of the host, and/or immunogenicity. Such endotoxoids may be prepared by a variety of chemical procedures (Johnson and Nowotny, 1964; Freedman and Sultzer, 1962;

NOWOTNY, 1963; MARTIN and MARCUS, 1966a, b; NOWOTNY, 1964), by ionizing radiation (PREVITE et al., 1967), or by combining endotoxin with the corresponding O antibodies (NOWOTNY et al., 1965). NOWOTNY (1968) reported that endotoxoid remains immunogenic in rabbits. It is evident from the foregoing, then, that, although much has been learned of endotoxicity in the last few years, the immunologic aspects of both toxic manifestations and tolerance require further investigations.

Endotoxins as Adjuvants

Extensive studies during the past decade have unequivocally documented the adjuvant effect of endotoxins, and substantial progress has been made in the analysis of its mode of action. In retrospect it appears likely that this phenomenon was encountered decades ago, although for obvious reasons purified lipopolysaccharides could not be used at that time. Of course, it is important to keep in mind that lipopolysaccharides are not the only adjuvants. Most widely used in experimental animals is FREUND'S adjuvant; a monograph on this adjuvant was published by FINGER (1964), and the topic of adjuvants was presented at a recent symposium of the International Association of Microbiological Societies (REGAMEY et al., eds., 1967). It must be kept in mind that many substances possess the capacity to enhance the immune response, reference being made only to the adjuvancy of lipids and surface-active substances, without attempting a review of the large literature (cf. BARRIE and COOPER, 1964; DRESSER, 1961; FINNEGAN and DIENNA, 1953).

Half a century ago, BIELING (1919) showed that pre-treatment of rabbits with dysentery bacilli enhanced the production of agglutinins against the typhoid bacillus. This enhanced response was interpreted as due to non-specific anamnestic recovery. In passing, it may be mentioned that the author considered his findings to be contrary to EHRLICH'S side-chain theory. More than four decades ago, LEWIS and LOOMIS (1926) documented the adjuvancy of *Brucella* infection on the immune response to an unrelated antigenic stimulus. These authors coined the term of "allergic irritability". Enhancement of the immune response was described in 1924 by KHANOLKAR (1924) under the heading of "non-specific training of antibody production". The author showed that gram-negative bacilli, including *E. coli* and *Pseudomonas aeruginosa*, in contrast to staphylococcus and yeast, enhance antibody production to a *Salmonella* strain. Adjuvancy presumably due to endotoxin was studied as long as 30 years ago also by RAMON and ZOELLER (1926). Detailed studies on the adjuvancy effects of pertussis vaccine on the immunogenicity of diphtheria and tetanus toxoid were reported as early as 20 years ago (GREENBERG and FLEMING, 1947, 1948; FLEMING et al., 1948; BARR, 1956; FARAGÓ and PUSZTAI, 1949). MACLEAN et al. (1940) clearly showed that the amount of tetanus antitoxin produced by immunized human subjects was more than 5 times greater when tetanus antigen was given together with typhoid-paratyphoid A and B vaccine than when the former was administered alone. The modern era commenced with the report by CONDIE et al. (1955a) on the adjuvant effect of meningococcal endotoxin in rabbits immunized with bovine serum albumin, and the authors stated in the abstract that preliminary data suggest that very large amounts of endotoxin may suppress the immune response.

During the past 12 years the adjuvant effect of endotoxins has been well documented. The pertinent data are summarized in Table 1. Perusal of the table clearly shows that endotoxins exert adjuvant effects in varied animal species, including mouse, rat, rabbit, guinea pig, and chicken. It is clear, too, that the immune response is enhanced to both soluble and cellular antigens. Among the former are bovine gamma globulin, bovine serum albumin, human serum albumin, and diphtheria toxoid, and among the latter sheep erythrocytes and Rauscher virus. Worthy of mention is the fact that in one study enhancement of the immune response to a hapten has been observed (TARMINA et al., 1968). Adjuvancy has been documented by increase of titers of circulating antibodies, increase in the number of antibody-producing cells, accelerated antigen elimination, shortening lag period of the immune response, and/or longer persistence of circulating antibodies.

JOHNSON et al. (1956) showed that endotoxins do not enhance the immune response to polysaccharides, namely, *Pneumococcus* type III, *Pasteurella* (*Francisella*) *tularensis*, and Vi antigens. So far as this reviewer is aware, the possible adjuvant effects of endotoxin on the immune response to various polysaccharide antigens have not been pursued further, unless enhanced antibody formation to sheep erythrocytes is due to polysaccharide antibodies. Such studies appear to be particularly needed, in as much as it has been shown that endotoxins are capable of breaking immunologic tolerance to polysaccharides, as discussed below. It will be of interest, too, to ascertain whether endotoxins enhance the antibody response to nucleic acids, for antibodies against these antigens have now been unequivocally identified (PLESCIA and BRAUN, 1967).

Use has been made of the adjuvancy of lipopolysaccharides in the production of high titered antisera, for example, antiserum against Gc antigen of human serum (TUROWSKA and TUROWSKI, 1965) and antiserum against Rauscher virus (SIBAL et al., 1968). Of considerable interest, too, are the observations on adjuvancy of endotoxin during immunization with tissue antigen (DAVIES et al., 1963).

Information on the amounts of lipopolysaccharides or endotoxins used for the demonstration of enhanced antibody production is summarized in Table 2. It is evident that, on a weight basis, striking differences exist with regard to effective amounts in various animal species, the rabbit and the chicken being far more susceptible than the mouse and the guinea pig. Dose response curves, however, so far as this reviewer is aware, have not been established in comparative studies with several antigens used in varied animal species, and information on minimally effective doses of endotoxin is, therefore, not available for definitive assessment.

Table 1. *Adjuvancy of endotoxin*

Animal species	Antigens[a]	Endotoxins	Effect on immune response[b]	Authors
Mouse	BGG	*S. typhosa*	Enhancement	CAMPBELL et al. (1966)
Mouse (Balb)	BGG	*S. marcescens*	Enhancement and shortened induction period	MERRITT and JOHNSON (1963)
Mouse	Sheep erythrocytes	*S. marcescens*	Enhancement	FINGER et al. (1967)
Mouse	Sheep erythrocytes	*S. marcescens*	Increase in number of antibody producing cells	FREEDMAN et al. (1966)
Mouse	BGG	*S. typhosa*	Enhancement after transfer of spleen cells to x-irradiated host	COHEN et al. (1964)
Mouse	Sheep erythrocytes	Various endotoxins	Enhancement	FREEDMAN et al. (1967a)
Mouse	Sheep erythrocytes	*S. marcescens*	Accelerated appearance of plaque-forming spleen cells	FINGER et al. (1967)
Mouse	Sheep erythrocytes	*E. coli*	Enhancement	STUART and DAVIDSON (1964)
Mouse	Rauscher virus	*S. abortus equi*	Enhancement	SIBAL et al. (1968)
Rat	BGG	*S. typhosa*	Enhancement of primary response	PIERCE (1967a)
Rat	BGG	*S. typhimurium*	Enhanced primary 19 S antibody response	PIERCE (1967b, c)
Rat (splenectomized)	BGG	*S. typhimurium*	Enhanced primary and secondary 7 S antibody response	PIERCE (1967b, c)
Rat	Rat heart	*E. coli*	Enhancement	DAVIES et al. (1963)
Rabbit	BSA	Meningococcus (crude)	Enhancement	CONDIE et al. (1955b)
Rabbit	Ovalbumin, BSA, diphtheria toxoid, *B. pestis* protein	Various LPS	Enhancement	JOHNSON et al. (1956)
Rabbit	BSA	Meningococcus	Enhancement	CONDIE et al. (1955a)

Species	Antigen	Endotoxin	Effect	Reference
Rabbit	HSA	*S. typhosa*	Enhancement	Kind and Johnson (1959)
Rabbit	BGG	*S. abortus equi*	Enhancement	Langevoort et al. (1963)
Rabbit	Sheep erythrocyte stromata	*S. typhosa*	Enhancement	Pearlman et al. (1963)
Rabbit	Arsanilic acid (hapten)-egg albumin	*S. enteritidis*	Enhancement	Tarmina et al. (1968)
Rabbit	Diphtheria toxoid	*S. typhimurium*	Enhancement	Prochazka (1961)
Rabbit	Gc (human serum) antigen	*E. coli*	Enhanced antibody production	Turowska and Turowski (1965)
Rabbit	BSA	*S. abortus equi*	Formation of antibody in spleen accelerated and enhanced; antibody also in thymus	Spurgash and Sibal (1967)
Guinea pig	Diphtheria toxoid	*B. pertussis*, *E. coli*, *E. coli* lipoid A	Enhancement	Farthing (1961)
Guinea pig	Diphtheria toxoid	*B. pertussis*, *E. Coli*	Enhancement	Farthing and Holt (1962)
Guinea pig	Sheep erythrocytes	*E. coli*	No increase in percentage of antibody producing cells of lymph nodes; enhancement of circulating antibody titers	Kierszenbaum et al. (1967)
Guinea pig	DNP-BSA	*S. typhimurium*	Enhancement	Hill and Rowley (1967)
Chicken	BSA	*S. abortus equi*	Enhancement (shorter induction time, longer persistence)	Luecke and Sibal (1962)
Chicken	BSA	*S. abortus equi*	Enhancement produced by spleen cells transferred to embryos	Sibal (1961)

[a] Abbreviations: BGG = bovine gamma globulin; BSA = bovine serum albumin; HSA = human serum albumin.
[b] Enhancement = increased titers of antibodies.

Table 2. *Amounts of endotoxins effective as adjuvants in various animal species*

Animal species	Effective amounts of endotoxins (μg/kg)	Ineffective amounts of endotoxins (μg/kg)	Authors
Mouse	2,500		FARTHING et al. (1967)
Mouse	500—5,000	50	JOHNSON et al. (1956)
Mouse	1,200		SIBAL et al. (1968)
Rat	400		PIERCE (1967a, b, c)
Rat	100		DAVIES et al. (1963)
Rabbit	0.1 to 5		JOHNSON et al. (1956)
Rabbit	3		TARMINA (1968)
Guinea pig	1,000		KIERSZENBAUM et al. (1967)
Chicken	4		SIBAL (1961)

Table 3. *Adjuvancy of endotoxin and time relationship to antigen administration*

Animal species	Antigens[a]	Adjuvancy of endotoxin[b] observed	not observed	Authors
Mouse	SRC	−24 hr	+10 to 12 ds	FREEDMAN et al. (1967a)
Mouse (Balb)	BGG	up to −7 ds, up to +6 ds		MERRITT and JOHNSON (1963)
Rabbit	HSA	0, up to +48 hr	−6 to 24 hr	KIND and JOHNSON (1959)
Guinea pig	Diphtheria toxoid	0, +24 hr	−24 hr, +>24 hr	FARTHING and HOLT (1962)
Chicken	BSA	0, +24 hr	−24 hr	LUECKE and SIBAL (1962)
Rat	Heart tissue	0	+8, −8	DAVIES et al. (1963)

[a] Abbreviations: SRC = sheep erythrocytes; BSA = bovine serum albumin; HSA = human serum albumin; BGG = bovine gamma globulin.

[b] + = time after antigen injection; — time prior to antigen injection; 0 = simultaneous injection with antigen; hr = hours; ds = days.

Of particular interest are the observations regarding the time relationships between injection of antigen and of endotoxic adjuvant. Pertinent information is summarized in Table 3. There is no question that, for endotoxin to enhance the immune response, it has to be injected either together with the antigen or within critical periods of time before or after antigen. It is noteworthy that marked differences exist between animal species: in the guinea pig injection of endotoxin either 24 hours before or more than 24 hours after antigen injection is ineffective; in contrast, in Balb mice enhancement was documented when endotoxin was injected as long as 7 days before or 6 days after administration of antigen. Clearly, additional studies utilizing various antigens and animal species are needed, and the reasons for the observed differences remain to be elucidated.

Endotoxin exerts yet another type of adjuvant effect that is of particular significance, namely, it is capable of breaking the state of immunologic tolerance (unresponsiveness). BROOKE (1965) demonstrated that *Salmonella typhosa* endotoxin terminates unresponsiveness in 10 week-old CAF_1 mice injected with specific soluble substance of type III pneumococcus, this effect being documented both by demonstration of circulating antibodies and by resistance to challenge with viable pneumococci. Endotoxin in amounts of 400 µg was injected intraperitoneally, and the polysaccharide was administered in amounts of 25 to 200 µg. Of 64 mice which had received endotoxin at various times after pneumococcal antigen, 62 survived the challenge. In contrast, of the control mice, which had received endotoxin alone or pneumococcal polysaccharide alone, all died. Likewise, 59 out 64 mice that had been injected with endotoxin and antigen formed demonstrable circulating antibodies. This effect of endotoxin is observed only if it is injected after, but not prior to, the pneumococcal polysaccharide. As yet unexplained is the observation that the effect of endotoxin is less marked in C3H/HeJ male mice, 9 weeks of age. MARGHERITA and FRIEDMAN (1965) reported that *Serratia marcescens* endotoxin increased resistance to challenge with viable pneumococci in animals that had been immunologically paralyzed with high doses of the corresponding polysaccharide. Although the authors considered the possibility that endotoxin may have terminated the state of paralysis, they suggested that a specific humoral factor was not involved in the production of nonspecific resistance. In a later study, MARGHERITA and PATNODE (1968) showed that endotoxin (50 µg) terminates immunologic unresponsiveness produced by 500 µg of type VIII pneumococcal polysaccharide when the animals were challenged 20 hours after endotoxin administration with antigen in amounts of 0.5 µg. In this connection it is important to call attention to the observations of NEEPER and SEASTONE (1963) who showed that immunologic unresponsiveness can be reversed by nonspecific mechanisms; water-in-oil emulsion abolishes unresponsiveness in mice induced by pneumococcal polysaccharide. CLAMAN (1963) made the interesting observation that immunologic tolerance, which is induced in adult mice by a single injection of supernate from ultracentrifuged bovine gamma globulin, does not develop when endotoxin is given intravenously. Attention may be called also to the implications of the model proposed by TALMAGE and PEARLMAN (1963), suggesting a dual effect of antigen, one being specific and the other nonspecific. It is conceivable that the nonspecific effect of antigen may be replaced by unrelated substances. COHEN et al. (1964) documented the adjuvant effect of endotoxin on the immune response of unresponsive cells. These results suggest that termination of immunologic paralysis can be effected by endotoxin and that seemingly nonspecific resistance engendered by endotoxin in unresponsive mice may have, in fact, a specific immunologic basis. It will be of considerable interest to learn whether lipid A and endotoxoid in suitable amounts also are capable of terminating the state of immunologic unresponsiveness. Additional investigations may reveal also whether the mode of action of endotoxin as an adjuvant to protein antigens differs strikingly from that responsible for termination of immunologic unresponsiveness to polysaccharides.

Finally, endotoxins may affect the immunologic apparatus by causing an increase in the titers of "normal" antibodies and/or in the number of immunoglobulin-producing cells in the absence of a known antigenic stimulus. If the term "adjuvancy" connotes only enhancement of the immune response to an antigenic stimulus, then, this effect of endotoxins on the immunologic apparatus cannot be considered as due to adjuvancy. The pertinent data on the effects of endotoxins on the immunologic activity in the absence of known antigenic stimuli are summarized in Table 4. It can be seen that endotoxins from a variety of gram-negative bacteria are effective and that this activity has been documented in a variety of animal species, including mice, rabbits, guinea pigs, and pigs.

A challenging thesis, relating to this ability of endotoxin to initiate non-specific responses, has been proposed by BRAUN, YAJIMA and NAKANO (to be published), postulating that upon antigenic stimulation, processing by macrophages results in the formation of a complex of non-specific activator and antigenic determinant, which normally results in a subsequent activation of appropriate pre-existing stem-cells of antibody-forming populations, the activator being guided into the proper stem-cells by permeability-modifying, membrane-associated reactions between the antigenic determinant (of the complex) and antibody-like recognition sites on the lymphocytic stem-cells. It is assumed that the activator can normally not enter other stem-cells since permeability barriers prevent its entrance; the consequence is a specific response. However, in the presence of endotoxin, or in the presence of other known modifiers of membrane permeability, the activator can enter other stem-cells; this "misguidance" then results in the activation of "wrong" stem-cells, the consequence being a "non-specific" response. BRAUN et al. have furnished some experimental support for this conclusion by studying specific and non-specific responses to an antigen such as sheep red blood cells, given with or without endotoxin, to mice that were tolerant to the adjuvant effects of endotoxin.

From the foregoing, then, it is clear that endotoxins may influence the immunologic apparatus in at least three sets of circumstances. Endotoxins may (1) enhance the immune response to certain antigenic stimuli, (2) terminate immunologic unresponsiveness (tolerance), and (3) cause an increase in the activity of the immune system in the absence of known antigenic stimuli.

Regarding the modes of adjuvancy of lipopolysaccharides and of the numerous other materials that have served as enhancers of the immune response, the following mechanisms may be considered.

Table 4. *Effect of endotoxins on immune response in the absence of antigenic stimulus*

Animal species	Endotoxins	Effect on immune response	Authors
Mouse	Various LPS	Increase in number of sheep cell antibody producing spleen cells	HEUER and PERNIS (1964)
Mouse	Various LPS	Increase in titer of bactericidal antibodies and of antibody-producing splenic plaques	MICHAEL (1966)
Mouse	*S. dysenteriae*	Increase in cidal antibody titers against various enteric bacteria, noted between 3 and 96 hr after endotoxin	MICHAEL et al. (1961)
Mouse	*E. coli* LPS	Increase in *V. cholerae* opsonins	JENKIN and PALMER (1960)
Mouse	*Brucella abortus* endotoxin (Boivin)	Increase in hemolysin-producing spleen cells only after sensitization	FREEDMAN et al. (1968)
Mouse	*S. marcescens*	Moderate increase in number of pre-existing plaque forming spleen cells	FINGER et al. (1967)
Mouse	Various LPS	Increased opsonic activity	ROWLEY and TURNER (1964)
Mouse	*E. coli* LPS	Increased opsonic activity	ROWLEY (1960)
Rabbit	Endotoxin	Increase in sheep erythrocyte plaques after single dose. Differences in effects of second dose related to protein content of first dose	FREEDMAN et al. (1967b)
Guinea pig	*S. typhimurium* LPS	Increase in titer of rabbit erythrocyte antibodies	HILL and ROWLEY (1967)
Pig	Various LPS	Increased opsonic activity	ROWLEY and TURNER (1964)

(1) Adjuvants affect the antigenic stimulus, such as slow release of antigen from the site, protection of the antigen from possible destruction, or other factors that relate to absorption of the antigen.

(2) Adjuvants affect antigen-processing, such as mobilization of operative cells, facilitation of interaction of antigen with membranes, phagocytosis or pinocytosis, or alteration of the very intracellular processing events that operate as part of the immune response.

(3) Adjuvants affect antibody-producing cells by stimulation of proliferation, alteration of protein synthesis, etc.

(4) Adjuvants may affect cells that are not strictly part of the immunologic apparatus, for example, causing release of DNA breakdown products, which in turn affect the immune system.

(5) Adjuvants may affect feedback mechanisms operative in antibody production.

Obviously, endotoxins and other adjuvants may have more than one of the above mentioned effects.

A great deal of effort has been directed toward the elucidation of the mode of action of endotoxins as adjuvants.

Early investigations by CONDIE and GOOD (1956) suggested that adjuvancy is related to toxic properties of endotoxin, in as much as production of tolerance by repeated injections of endotoxin into rabbits abolishes the adjuvancy of this material. Similarly, FREEDMAN et al. (1966), studying the adjuvant effect of *Serratia marcescens* endotoxin on the immune response, measured by the Jerne plaque technique, to sheep erythrocytes in tolerant and nontolerant mice, found that endotoxin in tolerant animals fails to effect an increase in the number of hemolysin-producing cells of the spleen. When tolerance was broken by the injection of carbon, the capacity to react to endotoxin with an increase in the number of antibody-forming cells was restored. However, a series of observations suggest that adjuvancy does not depend entirely upon the toxicity of endotoxins. This is evident from the observation of LUECKE and SIBAL (1962) who reported that, in chickens, endotoxin in amounts as small as 5 or 10 μg/kg sufficed to enhance the immune response to bovine serum albumin. These amounts are similar to those effective in the rabbit, although the latter is substantially more susceptible to pyrogenic and other toxic effects than the chicken. This conclusion is supported also by the findings that it is possible to reduce the toxicity of endotoxins without parallel reduction in adjuvancy. RAUSS et al. (1958), for example, showed that treatment of endotoxin with $Al(OH)_3$ diminishes toxicity and yet increases adjuvancy approximately 10 times. Also, lipid A, which is less toxic than the parent lipopolysaccharide was found to enhance immunogenicity, although it was somewhat less effective than the endotoxin (FARTHING, 1961). Of particular significance are the results with endotoxoids. JOHNSON and NOWOTNY (1964) reported that endotoxoids are capable of enhancing the antibody response of mice to serum proteins (human gamma globulin, bovine gamma globulin) and also noted that adjuvancy differs slightly depending upon the method used for detoxification of endotoxin.

Much remains to be learned regarding the adjuvancy of various endotoxin preparations. TARMINA et al. (1968) observed that *Salmonella enteritidis* endotoxin with adjuvant properties can be dissociated into molecules of lower molecular weight by means of sodium deoxycholate. The latter molecular species exert less adjuvancy than the former, suggesting that the colloidal properties of endotoxin are of import in its biologic activities. FREEDMAN et al. (1967a) studied the adjuvant effect on antibody-forming spleen cells in mice after pretreatment with *S. abortus equi* endotoxin. It is significant that differences were observed between endotoxins containing protein and a protein-free preparation and also whether or not endotoxin of the same source had been given previously. In connection with the above observations it is of interest to mention the findings of MILNER and FINKELSTEIN (1966) to the effect that the state of dispersion of endotoxin does not affect lethality in mice but has a profound influence on pyrogenicity and lethality in chick embryos. These observations were made by comparing clean cell walls of enterobacteriaceae with the respective endotoxins. Although it is not entirely clear why certain antigens are immunogenic and others are not, such as bovine gamma G protein and bovine gamma A protein, such differences must be related to those parts of the molecule that endow the latter with adjuvanticity rather than to the antigenic determinant. DRESSER (1968) made the interesting observation that vitamin A may act as adjuvant, probably because of its effects in damaging lysosomal membranes and thus stimulating cell division, and it is conceivable that endotoxin produces adjuvancy, in part at least, by a similar mechanism.

Particularly convincing are the studies suggesting that adjuvancy of endotoxins is related to the liberation of tissues breakdown products capable of increasing the immune response (cf. book on Nuclein Acids and Immunology, PLESCIA and BRAUN, 1968). BRAUN and NAKANO (1965) showed that DNA breakdown products, administered together with antigen, increase the number of antibody-forming cells present in the spleen of mice. This adjuvant effect is observed only when antigen is administered concurrently, unless permeability changes are effected, for example, by chlorpromazine. It should be noted that these oligodeoxyribonucleotides differ in their effect on the immune apparatus from endotoxin in as much as the latter stimulates antibody formation in the absence of an antigenic stimulus. MERRITT and JOHNSON (1965) as well as JAROSLOW (1960), also suggest that it is true cell injury and the liberation of nucleic acid products that account for adjuvancy of endotoxin. This concept is in accord with the findings of TALIAFERRO and JAROSLOW on the restoration of antibody formation in x-irradiated rabbits by nucleic acid derivatives (TALIAFERRO and JAROSLOW, 1960). Release of these nucleic acid breakdown products may be mediated by cell-associated antibodies (KESSEL et al., 1964). BRAUN (1964) made the interesting observation that degradation products of nucleic acid enhance the adjuvant effect of endotoxin. He found that when subeffective amounts of *Serratia marcescens* endotoxin are combined with nonpyrogenic DNA degradation products, the number of hemolysis-forming spleen cells increases to a greater extent than is effected by either material alone. DNA breakdown products (oligodeoxyribonucleotides) also are capable of stimulating antibody production in endotoxin-tolerant animals. It is noteworthy, too, that endotoxin-tolerant animals react in a normal fashion to the injection of antigens, indicating that tolerance does not preclude a normal immune response (FREEDMAN et al., 1966).

Studies on the comparative efficacy in various animal species of endotoxins and endotoxoids regarding cell injury, release of DNA breakdown products, and adjuvancy are indicated in order to clarify further the mode(s) of action of these adjuvants.

FREEDMAN and BRAUN (1965) observed that oligodeoxyribonucleotides stimulate the phagocytic activity of the reticuloendothelial system. Nucleic acid breakdown products also may play a role in antigen-processing by macrophages (JOHNSON and JOHNSON, 1968). These authors showed that macrophages exposed to BGG and poly A-U, when re-injected into mice, enhance the antibody response when comparison is made with animals that received antigen and poly A-U but no macrophages. In this connection, attention may be called to the recent studies by ARGYRIS (1967), NICOL et al. (1966), FREI et al. (1965), FISHMAN and ADLER (1967), UNANUE and ASKONAS (1968), BISHOP et al. (1967), as well as GALLILY and FELDMAN (1967) dealing with the role of macrophages in the immune response. It should be kept in mind, however, that intracellular residence in polymorphonuclear leucocytes and alveolar macrophages may result in destruction of immunogenicity of *E. coli* in contrast to that in peritoneal macrophages (COHN, 1964). MOSIER (1967) concludes that two cell types are required for antibody production and that these cell types can be separated by their ability or inability to adhere to plastic dishes. It is of further interest to note that brief exposure of the adherent cells to antigen suffices for initiation of the *in vitro* antibody response. Clearly, the effects of endotoxins in these systems deserve investigation.

In view of the fundamental role of the thymus in immunologic competence of certain animal species (GOOD and GABRIELSEN, Eds., 1964; METCALF, 1967), the question arises as to whether endotoxin's adjuvancy is related to the thymus. CAMPBELL et al. (1965, 1966) studied the effects of thymectomy in CBA mice at either 2 or 5—7 weeks of age; bovine gamma globulin was used as antigen and *S. typhosa* lipopolysaccharide as endotoxin. The immune response was measured by both antigen elimination and hemagglutinin titration. Endotoxin proved to be effective irrespective whether thymectomy or sham-thymectomy had been carried out in the older mice. Reduction in antibody response was observed, however, in mice that had been thymectomized at two weeks of age. It is evident, therefore, that the thymus is not required for endotoxin to be effective as an ajduvant. The reduced immune response in the animals thymectomized shortly after birth may reflect the absence of a full component of immunologically potent cells. In this connection it is of interest to note that endotoxin causes loss of thymic lymphocytes and alteration of thymic morphology in the mouse, as well as temporary stunted growth, this temporary effect seemingly related to the development of endotoxin tolerance (KIND et al., 1967).

ROWLANDS et al. (1965) studied the effect of endotoxin on the thymus in CBA female mice, 5 to 6 weeks old. Injection of *S. typhi* endotoxin caused a decrease in weight of the thymus during the first 3 days; the weight returned to normal by the 10th day. Histologically, the most striking effect of endotoxin was the rapid loss of thymic lymphocytes associated with an apparent increase in cells with pyroninophilic cytoplasm and an increase in the RNA/DNA ratio. The authors suggest that the adjuvant effect of endotoxin may be related to its effect on the thymus.

Some of the future studies may utilize systems described during the past few years. Although thymus cells apparently do not produce antibody, cells from a thymus graft may proliferate in response to certain antigenic stimuli. These cells have been referred to as reactor cells. GERSHON et al. (1968) made the interesting observation that antigen exposure prior to transfer may interfere with the activity of the reactor cells and this inactivity may contribute to immunologic tolerance. LANDY et al. (1965) reported that *S. enteritidis* O antigen injected into rabbits in amount of 5 μg resulted in an increase in the number of O antibody-forming cells in the thymus. This reaction was associated with loss of weight and cellularity of this organ. It is possible that differences exist in this reactivity in different animal species. OSOBA (1968) showed that thymectomized mice, heavily irradiated and given isogenic bone marrow, failed to recover the capacity to produce normal numbers of antibody-forming cells after antigenic challenge. Implantation of cell-impermeable diffusion chambers containing thymus restored this capacity. The author suggests that the influence of the thymus is directed at the differentiation of antigen-insensitive precursor cells in bone marrow to antigen-sensitive cells, and concludes that the thymic influence is mediated by a humoral factor. Recent evidence (CLAMAN et al., 1968) indicates that thymus and bone marrow cells together play a role in immunocompetence. It will be of interest to determine the effects of lipopolysaccharide under these experimental conditions, particularly since endotoxin affects both bone marrow and the lymphatic system.

A number of studies, bearing on the mode of action of endotoxin as adjuvant, were carried out in animals whose immunologic apparatus had been altered by splenectomy, irradiation, corticoids, or immunosuppressive drugs. PIERCE (1967b, c) observed that in intact rats *S. typhi* endotoxin enhances the primary immune response to BGG, and the antibodies produced are of the 19S class. In splenectomized animals endotoxin also acted as adjuvant, but both in the primary and secondary immune response the antibodies produced were of the 7S variety. On the basis of these observations the author suggests that adjuvancy is due to interference with feedback inhibition of antibody production and this effect results in enhanced and prolonged antibody production. This assumption is supported by the observation that passive immunization of splenectomized rats suppressed the primary response and also interfered with sensitization for a secondary response. In the intact rat passive immunization interfered with sensitization for a secondary response. Endotoxin eliminates the suppressive effect of passive immunization. Attention may be called to the observations of SOLLIDAY et al. (1967) that *Salmonella typhi* endotoxin overcomes the immunosuppressive effect of antibodies directed against the immunizing antigen. The role of feedback in antibody production is in accord with current ideas, but it remains to be determined whether, and to what extent, adjuvancy is due to this mechanism. Similar studies utilizing other animal species and additional antigens are needed.

Preliminary experiments by KIND and JOHNSON (1959) indicate that endotoxin may partially restore antibody formation in x-irradiated rabbits. Similarly, COHEN et al. (1964) observed that endotoxin acts as adjuvant in x-irradiated mice after transfer of spleen cells and antigenic stimulation. That endotoxin counteracts cortisone-induced immuno-suppression was reported by WARD and JOHNSON (1959). It is likely that both compounds affect the early events of the immune response. Of interest are the observations by KIERSZENBAUM et al. (1967) who reported that diminution of proliferative reactions by colchicine did not abolish the adjuvant effect of lipopolysaccharide. It is on the basis of these observations that the authors suggest that activation of antibody production by cells rather than cell differentiation or hyperplasic phenomena account for adjuvancy. FOX et al. (1968) confirmed the observation that endotoxin causes an increase in the number of pre-existing antibody forming cells and that, as revealed by analysis of the effects of FUDR, this effect does not follow the same pathway as that stimulated by antigen.

There is no question that endotoxin acts as an adjuvant in the primary immune response to protein antigens (JOHNSON, 1964). There is less agreement regarding its effect on the secondary response. RÉTHY (1967) who carried out extensive studies on adjuvancy, concluded that the secondary response is enhanced by these macromolecules. PIERCE (1967a) failed to observe enhancement of the secondary response by *S. typhosa* endotoxin in rats immunized with bovine gamma globulin. Antibodies produced in either the primary or secondary response after immunization with antigen alone had no detectable precipitating activity. In contrast, precipitating 7S antibodies appeared during the secondary response in animals injected twice with antigen and endotoxin. From these observations it is evident that additional studies in various animal species and with different antigens are needed regarding the adjuvant effects of endotoxin on the secondary response, and that careful attention must be paid to the properties and class of antibodies produced under these conditions. In this connection attention may be called to the observation (SASSEN et al., 1968) that heat-treated, but not untreated, human gamma globulin is a good antigen for the initiation of the primary antibody response in mice. Both materials, in contrast, are effective as immunogens for the secondary response. This observation suggests important differences between primed and non-primed spleen cells in their requirements of cellular organization for the initiation of the antibody response.

The adjuvancy of endotoxin can be investigated and analyzed also in cell transfer experiments. McKenna and Stevens (1957) as well as Stevens and McKenna (1958) studied the production of antibodies by spleen cells *in vitro*. Endotoxin administered 24 hours before antigen increased the antibody response. They reported also that endotoxin was effective *in vitro* in enabling spleen fragments to produce antibodies. Kong and Johnson (1963) repeated these experiments and were unable to confirm the results and the conclusions. Their studies indicate that in the cell transfer system the adjuvant effect of endotoxin is at best suggestive. It remains to be seen whether such effects can be documented if it were possible to maintain cell viability of the tissues for longer periods of time. Sibal (1961) in elegant experiments, has shown that endotoxin given to adult chicken together with BSA as antigen led to antibody production by spleen cells transferred to the chick chorioallantoic membrane. Cells transferred 24 hours after antigen administration produced antibodies in only 2 out of 59 transplants. In contrast, spleen cells taken 24 hours after administration of endotoxin (*S. abortus equi*) and antigen consistently produced antibody following transfer to embryonic recipients.

Pearlman et al. (1963) studied the question whether adjuvancy might be mediated by action of endotoxin on complement. Previously, Talmage and Pearlman (1963) had postulated that complement may act as inhibitor of cell divisions and complement fixation or inactivation within antibody-forming cells may be a necessary step in the immune response. Since complement titration of intracellular material is not feasible as yet, Pearlman et al. (1963) studied its effect on serum complement. Both *S. typhi* and *S. enteritidis* lipopolysaccharides were used. In albino rabbits, endotoxin produced a striking decrease in complement activity within 6 hours, complement titers returning to normal levels by 24 hours. In some instances complement titers increased beyond the pre-injection levels for at least 48 hours thereafter. On the basis of these results the authors postulate that endotoxin may serve as adjuvant because of the effects on serum complement activity. These results are in accord with the findings from Mergenhagen's laboratory (Gewurz et al., 1968). If, in fact, adjuvancy of lipopolysaccharide is related to its effect on the complement system, then, the interaction of endotoxin and cell membranes becomes of interest. Spielvogel (1967), by means of electron microscopy, has provided evidence that endotoxin adheres to platelets and that this interaction may require the participation of complement, in addition to antibody. In connection with the possible role of complement in adjuvancy it is of interest to note that, according to Rosse et al. (1968), partially purified component (C'la) increases the binding of cold-reacting antibodies to appropriate cells. It will be of interest to learn whether endotoxin affects this reaction. However, it is not likely that the ability of endotoxin to fix complement is the major effect responsible for adjuvancy, for endotoxoid enhances antibody formation and is less active in complement interaction.

Brief reference will be made to some miscellaneous studies related to the mode of action of endotoxins as adjuvants, studies that need to be expanded in the future. STERZL et al. (1961) made the observation that lipopolysaccharide of *S. typhi* or *E. coli* is less effective as an adjuvant in very young than in adult rabbits. The difference may conceivably be related to lack of sensitization to endotoxin of the young animal. In this connection it is of interest to note that *Corynebacterium parvum* is a powerful adjuvant in adult rabbits but not in neonatal animals (PINCKARD et al., 1967). Studying the mechanism of adjuvancy of endotoxin, the Soviet investigators UCHITEL and KHASMAN (1965) found that doses of *S. typhi* endotoxin which stimulate antibody production in chinchilla rabbits also cause an increase in the incorporation of labeled methionine-S^{35} into blood serum proteins, lymphnodes, spleen, and adrenal glands. It is postulated by the authors that adjuvancy is related to increased protein synthesis. It will be of interest to determine whether endotoxoid has the same effect. ROSENBLATT and JOHNSON (1963) observed that the adrenergic blocking agent, dibenzyline, does not inhibit adjuvancy, nor does epinephrine given intravenously potentiate adjuvancy. Thus, no evidence was obtained on the role of the autonomic nervous system as a critical factor in adjuvancy. Having established that endotoxin counteracts the immunosuppressive effect of cortisone, WARD et al. (1961) studied the morphologic alterations associated with this treatment. The primary antibody response to bovine gamma globulin in the rabbit stimulated by *E. coli* endotoxin was associated with the appearance and proliferation of pyroninophilic reticular cells in the centers of splenic follicles. The same authors (WARD et al., 1959) concluded from the results of another study that endotoxin augments the histologically demonstrable changes observed during the primary immune response. The morphologic changes associated with the antibody response and as affected by endotoxin were described also by THORBECKE et al. (1962). FARTHING and HOLT (1962) concluded from their studies that hyperplasia of antibody-producing cells is an important factor in adjuvancy. A study in rabbits by LANGEVOORT et al. (1963) revealed that administration of endotoxin together with BGG not only enhanced serum antibody titers but also plasmacellular reaction and the formation of secondary nodules arising in the white pulp of the spleen.

From a comparative study of the effects of various lipids (triolein, cholesterol oleate, ethyl palmitate) and of endotoxin, STUART and DAVIDSON (1964) emphasize the role of the macrophage in the production of antibodies to foreign cells. DAVIES et al. (1963) studied the effect of lipopolysaccharide on the immune response to homologous heart tissue homogenate in rats and observed a marked adjuvant effect. This adjuvant effect was present only when antigen and lipopolysaccharide were injected together rather than separately. The authors postulate that the auto-immune process, at least in part, results from modification of the tissue antigen by linkage to endotoxin.

A definitive appraisal of the mode of action of endotoxin as an adjuvant is rendered difficult because of two facts, namely, the numerous alterations produced by these macromolecules in the host and our inadequate knowledge of the major events of the early immune response. Because of the numerous biologic alterations produced by endotoxin in susceptible hosts, it is difficult to relate, on a cause-and-effect basis, those events that are concerned with enhancement of the immune response. For example, increased protein synthesis produced by endotoxin may or may not play the major role in adjuvancy. Several approaches may reveal additional information, particularly if the experiments compare endotoxin, lipid A, and endotoxoids, and utilize different animal species, as well as immunologically virgin, "allergic", and immune animals, To the reviewer it seems particularly timely to study the effects of endotoxin with the macrophage RNA transfer system, just described by PINCHUCK et al. (1968). Following previous studies on the effects of RNA extracted from macrophages after exposure *in vitro* to antigen on the elicitation of the immune response of lymph node cells from non-immunized animals, these authors made the important observation that antibody formation is initiated in poorly responsive normal mice following transfer of the RNA extract from macrophages that had been incubated with synthetic polypeptides. Thus, it should be possible to determine whether endotoxin is operative as an adjuvant in the emergence of immunologically operative RNA and/or of immunocytes following exposure to this RNA-antigen product. Also, the model of HARRIS and CRAMP (1968) and of HARRIS (1968) on DNA synthesis measured by uptake of (^{14}C)-thymidine by spleen cells of rabbits immunized with protein antigens and the effect of antigen, antibody, and antigen-antibody complexes, appears to lend itself to further study of the adjuvancy of endotoxin.

Endotoxins as Immunosuppressants

As early as 1955, CONDIE et al. (1955a) reported that, on the basis of preliminary experiments, very large amounts of endotoxin, instead of increasing the immune response of rabbits to bovine serum albumin, suppressed antibody production. Surprisingly little attention has been given to endotoxins as immunosuppressants during the ensuing decade. BRADLEY and WATSON (1964) reported that endotoxin administered to mice either with actinophage MSP8 or three days after antigen inhibited the production of neutralizing antibodies. This effect was noted with large doses of endotoxin (4,000 μg/kg). Daily administration of endotoxin in these amounts after antigenic stimulation also caused a marked reduction in the production of antibodies. Discontinuation of endotoxin administration was followed, after a lag period, by the formation of antibodies. The secondary response was also affected. From these experiments it is evident that endotoxin only temporarily decreased or suppressed the immune response.

A different and seemingly novel type of immunosuppression came to light during studies of a bacterial antigen common to numerous enterobacteriaceae. The pertinent information leading to these studies is briefly summarized herewith. In 1962, KUNIN et al. (1962) observed that erythrocytes coated with crude preparations of O antigens of various serogroups of *E. coli* and *S. typhosa* were agglutinated by *E. coli* O14 but not by other group-specific *E. coli* antisera. Even more surprising was the observation that red blood cells coated with O antigen preparations from other enteric bacteria, such as strains of *Proteus, Shigella, Salmonella*, and *Klebsiella*, also were agglutinated by *E. coli* O14 antiserum, although the degree of hemagglutination varied. On the basis of these observations the authors suggested that various enterobacteriaceae contain a hitherto overlooked common hapten. Since antisera obtained by immunization of rabbits with various other enterobacteriaceae do not contain hemagglutinating antibodies against this common antigen in high titers, they suggested that the antigen is non-immunogenic (haptenic) in enteric bacteria other than *E. coli* O14 and fully immunogenic in the latter. For the sake of brevity, this common antigen of enterobacteriaceae will be referred to as CA. Attempts by KUNIN (1963) to purify CA were only partially successful, the isolated product, composed of polysaccharide and polypeptide, no longer coating erythrocytes nor inducing a significant immune response. CA was detected also in spheroplasts of *Salmonella typhosa* (WHANG and NETER, 1965a). Antibodies against CA of rabbit origin are largely present in the 19S fraction, irrespective as to whether early or late antisera were studied (WHANG et al., 1967). Antibodies against CA can be demonstrated by hemagglutination and its hemolytic modification. As yet unexplained is the fact that these antibodies fail to agglutinate heterologous enteric bacteria containing CA, although this antigen is present on the surface, as revealed by fluorescent antibody studies (AOKI et al., 1966). Nor do CA antibodies precipitate the antigen. It is of more than passing interest that the antibodies do opsonize various enterobacteriaceae for phagocytosis by rabbit peritoneal polymorphonuclear leucocytes (DOMINGUE and NETER, 1966a). DOMINGUE and NETER (1966b) also observed that latex particles modified with CA are opsonized in the presence of CA antibodies. Lipopolysaccharide and its lipid A component interfere with opsonization by CA antibodies, in spite of the fact that latex particles treated with both antigens absorb CA antibodies equally well as do particles treated with CA alone. These observations suggest the possibility that CA antibodies may protect against infection by various enterobacteriaceae. Indeed, preliminary studies have revealed that this approach, by both active immunization with CA and passive immunization with CA antibodies, merits further exploration (GORZYNSKI and NETER, 1968).

Since *E. coli* O14 produces more CA than other enterobacteriaceae, the possibility was considered that the differences in immunogenicity of CA preparations may be only quantitative rather than qualitative. NETER et al. (1964), however, showed that intravenous injection of equivalent amounts of CA of *E. coli* O14 and of other enterobacteriaceae caused a specific immune response in the rabbit only with the former antigen. It was necessary, therefore, to search for another explanation regarding the striking differences in immunogenicity of CA from various sources. In order to determine whether CA is an integral part of the lipopolysaccharide (endotoxin) macromolecule, WHANG and NETER (1962) carried out a comparative study of CA in crude antigens (supernates of agar-grown cultures) and highly purified lipopolysaccharides. The antigens were used for modification of erythrocytes and then tested with *E. coli* O14 and the corresponding O antisera. While the latter antisera produced the expected agglutination of erythrocytes modified with either crude or purified antigens, *E. coli* O14 antiserum caused agglutination only of red blood cells modified with the crude antigens, with the single exception of a *S. sonnei* lipopolysaccharide. The lipopolysaccharides used in these studies were prepared according to WESTPHAL'S phenol-water extraction method and kindly supplied by Professor WESTPHAL and Dr. LÜDERITZ. In view of these findings, unsuccessful attempts were then made to detect CA in the discarded fractions obtained during isolation and purification. Finally, it was learned that the cultures had been killed with ethanol prior to extraction. SUZUKI et al. (1964a), then, found that CA is ethanol-soluble and can be separated from the ethanol-insoluble O antigen (lipopolysaccharide). Thus, ethanol treatment of the cultures presumably had resulted in the elimination of CA. It may be noted in passing that the isolated CA exhibits only minimal toxicity characteristic of endotoxin, and this minimal toxicity may have been due to contamination with small amounts of endotoxin (KESSEL et al., 1966d). Surprising was the observation that the ethanol-soluble CA fraction from organisms other than *E. coli* O14 proved to be highly immunogenic upon intravenous injection into the rabbit. Injection of the mixture of the ethanol-soluble and insoluble fractions resulted in a minimal antibody response to CA. Thus, it was shown that CA produced by organisms other than *E. coli* O14 is, in fact, highly immunogenic and that this immunogenicity is decreased by the simultaneously present ethanol-insoluble fraction. It appeared, therefore, that CA as produced by enterobacteriaceae other than *E. coli* O14 represents the counterpart of haptens: both react *in vitro* with the corresponding antibody; hapten, however, is non-immunogenic unless coupled to a carrier (Schlepper), whereas CA is immunogenic and its immunogenicity is decreased by the simultaneous presence of the ethanol-insoluble fraction. Subsequent studies have revealed that purified lipopolysaccharides, when mixed with CA prior to immunization, inhibit the immune response of the rabbit to CA (SUZUKI et al., 1964b). Thus, lipopolysaccharide was identified as an immunosuppressant. It is important to point out that lipopolysaccharides do not affect the antigenic determinant by hindrance or other mechanisms, since CA reacts with antibodies equally well *in vitro* irrespective as to whether lipopolysaccharide is present or absent. Equally important is the observation that lipopolysaccharide when injected separately, albeit simultaneously, does not inhibit CA antibody production. On the basis of these observations it was postulated that CA, being fully immunogenic in the isolated state, is complexed with the immuno-suppressant lipopolysaccharide (endotoxin), thus accounting for the lack of a specific immune response of rabbits injected with crude supernates or bacterial suspensions other than *E. coli* O14. Further studies of the immunosuppressive effects of endotoxin revealed the following. Lipopolysaccharide suppresses the immune response of the rabbit to CA even when the mixtures are injected repeatedly. In one such experiment the antibody titers of rabbits so immunized increased from <1:10 to 1:14, at best an insignificant response. In contrast, injection of identical amounts of CA in the absence of lipopolysaccharide resulted in a titer of 1:108,800 (WHANG and NETER, 1967b). In addition, lipopolysaccharide-CA mixtures do not elicit a secondary response in animals primed to CA. suggesting that the inhibitor prevents the specific stimulation by antigen of memory cells leading to the emergence of antibody-producing cells. It is of considerable interest to note that lipid A has similar immunosuppressive effects (WHANG et al., 1965). Although lipopolysaccharide effectively inhibits the production of circulating CA antibodies, rabbits immunized with mixtures of CA and lipopolysaccharide are immunologically primed and respond with an accelerated and more intense production of CA antibodies upon the injection of subeffective amounts of isolated CA (NETER et al., 1966). The question was entertained whether endotoxin permits the emergence of committed or memory cells in the absence of antibody-excreting cells or whether liberation of antibody from the latter population is prevented. To this end, DOMINGUE and NETER (1967) utilized the Jerne plaque technique and observed that the number of CA antibody-producing cells of spleen or lymph nodes is not significantly increased upon immunization of rabbits with mixtures of CA and lipopolysaccharide. In contrast, a striking increase in the number of plaque-producing cells occurs upon injection of isolated CA.

That the immunosuppressive effect of lipopolysaccharide and its lipid A component is not restricted to one particular antigen is evident from the observation that these materials also suppress the production by rabbits of antibodies directed against the common antigen of gram-positive bacteria (WHANG and NETER, 1967a). This particular antigen was first described by RANTZ et al. (1956). It is noteworthy that in this system, too, immunosuppression is observed only when antigen and inhibitor are injected as a mixture and not when injected separately, albeit simultaneously.

Several possibilities were considered to explain the above observations. (1) Lipopolysaccharide may be a competing antigen. This assumption appears to be untenable because separate injection of antigen and endotoxin does not result in immunosuppression and because lipid A, at best, is a poor immunogen. (2) Immunosuppression is the result of tolerance. This thesis is incompatible with the observation that animals immunized with mixtures of antigen and lipopolysaccharide are immunologically primed. (3) Immunosuppression is due to toxicity of lipopolysaccharide. This explanation does not appear to be reasonable, for there is no basis for the assumption that toxicity is dependent upon simultaneous injection of endotoxin and antigen and because lipid A, of lesser toxicity than endotoxin, produces the same effect.

On the basis of the above observations the following working hypothesis is suggested. It is assumed that lipopolysaccharide or its lipid A component interact with the antigen and that this reaction assures the presence of the inhibitor on or in cells involved in the early phases of the immune response, perhaps by altering the uptake or processing of the antigen. The fact that separate injection of antigen and endotoxin does not suffice to effect immunosuppression may then be explained by the assumption that under these conditions the inhibitor may not be present in the majority of cells involved in antigen-uptake or processing. If this assumption were correct, one may consider future studies on immuno-suppressants that are coupled, perhaps like hapten and Schlepper, to effect immunosuppression under conditions in which separate injection of immunosuppressant and antigen might be ineffective. With these considerations in mind, a study of the effects of other lipid substances as possible inhibitors of the immune response was undertaken, and these investigations are facilitated by the fact that lipopolysaccharides readily complex with a variety of other materials (NETER et al., 1953; NETER et al., 1955; NETER et al., 1958a; LÜDERITZ et al., 1958; NETER et al., 1958b). Indeed, cardiolipin a naturally occurring cell component, proved to be such an inhibitor of the immune response of the rabbit to both common antigens (WHANG and NETER, 1968a). Again, inhibition takes place only when cardiolipin is injected together with the antigen.

Attention was then focused on the reasons for the immunogenicity of CA produced by *E. coli* O14. SUZUKI et al. (1966) reported that CA of *E. coli* O14 is ethanol-insoluble and thus differs from CA of identical specificity produced by other enterobacteriaceae. Further, a product of *Pseudomonas aeruginosa*, presumably an enzyme, destroys CA of enteric bacteria other than *E. coli* O14 but not that of the latter (WHANG and NETER, 1964). A similar product was isolated from psychrophilic *Pseudomonas* (WHANG and NETER, 1965b). On the basis of these observations, then, it was suggested that the identical antigenic determinant exists in two different molecular forms. Unfortunately, the chemical nature of these molecules and of the antigenic determinant accounting for CA specificity have not yet been elucidated.

Antibodies against the common antigen of gram-negative bacteria are present in serum of healthy subjects and in commercial gamma globulin preparations (WHANG and NETER, 1963). The antigenic stimulus conceivably could be due to colonization with either *E. coli* O14 or other CA-producing enteric bacteria. If the latter were the case, to explain CA antibody production in spite of the presence of immunosuppressant lipopolysaccharide, at least two possibilities present themselves: either the presumed CA-lipopolysaccharide complex is split, possibly by microbial or tissue enzymes, or the immunosuppressant effect of lipopolysaccharides is countered by antagonists. The latter possibility is suggested by the observation that cholesterol inhibits the immunosuppressive activity of lipopolysaccharides (WHANG and NETER, 1968b). Studies along these lines are needed to account for moderate CA immune response of rabbits to the injection into the footpad of supernates of enteric bacteria other than *E. coli* O14, with or without Freund's adjuvant (GORZYNSKI et al., 1963) and for the enhanced immune response to the injection of erythrocytes modified with both CA and lipopolysaccharide in the cell-attached rather than soluble states (NETER et al., 1964). The observation by DIAZ and NETER (1968) that children with shigellosis respond with the production of CA antibodies significantly better than patients with salmonellosis or urinary tract infection suggests that the microbial flora of the colon may affect the presumed complex.

The above mentioned investigations, then, have led to the recognition of two seemingly different models of immune suppression by endotoxin: In one, a systemic effect accounts for temporary inhibition of antibody production, and in the other, lipopolysaccharide is carried presumably with the antigen to the cells involved in the immune response. The latter mechanism may be the counterpart of hapten and carrier (Schlepper). In the schema of immune suppression presented below endotoxin represents types II(a) and II(c).

Mechanisms of Immune Suppression

I. Immunologically produced immune suppression
   (a) By antigens: tolerance, unresponsiveness
   (b) By antibodies: feedback inhibition

II. Non-immunologically produced immune suppression
   (a) Mechanisms not restricted to immunologic apparatus, such as irradiation, immunosuppressive drugs, corticoids, endotoxin, viruses
   (b) Mechanisms largely restricted to immunologic apparatus, such as thymectomy and anti-lymphocytic serum (particularly if a more cell-specific antiserum can be produced)
   (c) Antigen-carried immunosuppressants, such as endotoxin and cardiolipin (counterpart of Schlepper in rendering hapten immunogenic)

Outlook

Although much has been learned of the immunologic aspects of endotoxins and their biologic effects, many moot questions remain. It can be expected that progress will be made in the future, leading to a more complete understanding of these remarkable bacterial products and of their extraordinarily complicated effects in susceptible hosts. Among the areas in need of additional research are the structure of the macromolecule and the identification of biologically active parts, the reasons for the striking differences in susceptibility to endotoxins of various animal species and tissues, the role and precise identification of immune mechanisms operative in endotoxicity, adjuvancy, and immunosuppression, and the identification and function of antibodies directed against parts of the endotoxin molecule other than the R and O antigenic determinants. These investigations, in turn, promise to add to our understanding of basic aspects of various biologic phenomena, including the complicated process of the immune response.

References

ADKINSON, N. F., L. M. LICHTENSTEIN, H. GEWURZ, and S. E. MERGENHAGEN: The generation of anaphylatoxin by interaction of endotoxin and guinea pig serum. Fed. Proc. **27**, 316 (1968).

AOKI, S., M. MERKEL, and W. R. MCCABE: Immunofluorescent demonstration of the common enterobacterial antigen. Proc. Soc. exp. Biol. (N.Y.) **121**, 230—234 (1966).

ARGENTON, H., H. BECKER, H. FISCHER, J. OTTO, R. THIEL u. O. WESTPHAL: Über die resistenzsteigernde Wirkung von Lipoid A beim Menschen. Dtsch. med. Wschr. **86**, 774—781 (1961).

ARGYRIS, B. F.: Role of macrophages in antibody production. Immune response to sheep red blood cells. J. Immunol. **99**, 744—750 (1967).

ARREDONDO, M. I., and R. F. KAMPSCHMIDT: Effect of endotoxins on phagocytic activity of the reticuloendothelial system of the rat. Proc. Soc. exp. Biol. (N.Y.) **112**, 78—81 (1963).

ATKINS, E., and E. S. SNELL: Fever. In: The inflammatory process. New York: Academic Press 1965.

BARR, M.: The effect of typhoid-paratyphoid A and B vaccine on the antigenic efficiency of tetanus toxoids of different quality. Brit. J. exp. Path. **37**, 11—19 (1956).

BARRIE, J. U., and G. N. COOPER: Enhancement of the primary antibody response to particulate antigens by simple lipids. J. Immunol. **92**, 529—539 (1964).

BENACERRAF, B., and M. M. SEBESTYEN: Effect of bacterial endotoxins on the reticuloendothelial system. Fed. Proc. **16**, 860—867 (1957).

BERCZI, I., L. BERTOK, and T. BEREZNAI: Comparative studies on the toxicity of *Escherichia coli* lipopolysaccharide endotoxin in various animal species. Canad. J. Microbiol. **12**, 1070—1071 (1966).

BERGER, F. M., and G. M. FUKUI: Endotoxin induced resistance to infections and tolerance. Proc. Soc. exp. Biol. (N.Y.) **114**, 780—783 (1963).

BERRY, L. J., M. K. AGARWAL, and I. S. SNYDER: Comparative effect of endotoxin and reticuloendothelial "blocking" colloids on selected inducible liver enzymes. In: The reticuloendothelial system and atherosclerosis. New York: Plenum Press 1967.

— D. S. SMYTHE, and L. G. YOUNG: Effects of bacterial endotoxin on metabolism. I. Carbohydrate depletion and the protective role of cortisone. J. exp. Med. **110**, 389—405 (1959).

BIELING, R.: Untersuchungen über die veränderte Agglutininbildung mit Ruhrbacillen vorbehandelter Kaninchen. Z. Immun.-Forsch. **28**, 246—279 (1919).

BISHOP, D. C., A. V. PISCIOTTA, and P. ABRAMOFF: Synthesis of normal and "immunogenic RNA" in peritoneal macrophage cells. J. Immunol. **99**, 751—758 (1967).

BLADEN, H. A., H. GEWURZ, and S. E. MERGENHAGEN: Interactions of the complement system with the surface and endotoxic lipopolysaccharide of *Veillonella alcalescens*. J. exp. Med. **125**, 767—786 (1967).

BOGGS, D. R., J. C. MARSH, P. A. CHERVENICK, G. E. CARTWRIGHT, and M. M. WINTROBE: Neutrophil releasing activity in plasma of normal human subjects injected with endotoxin. Proc. Soc. exp. Biol. (N.Y.) **127**, 689—693 (1968).

BRADLEY, S. G., and D. W. WATSON: Suppression by endotoxin of the immune response to actinophage in the mouse. Proc. Soc. exp. Biol. (N.Y.) **117**, 570—572 (1964).

BRAUN, W.: Influence of nucleic acid degradation products on antibody synthesis. In: Molecular and cellular basis of antibody formation. New York: Academic Press 1965.

BRAUN, W., and W. FIRSHEIN: Biodynamic effects of oligonucleotides. Bact. Rev. **31**, 83—94 (1967).

—, and M. NAKANO: Influence of oligodeoxyribonucleotides on early events in antibody formation. Proc. Soc. exp. Biol. (N.Y.) **119**, 701—707 (1965).

— Y. YAJIMA, and M. NAKANO: Influence of endotoxins on the non-specific initiation of antibody formation and its possible relation to altered lymphocyte permeability. To be published.

BROOKE, M. S.: Conversion of immunological paralysis to immunity by endotoxin. Nature (Lond.) **206**, 635—636 (1965).

CAMPBELL, P., P. KIND, and D. T. ROWLANDS, JR.: Relation of the effects of endotoxin on the thymus and on antibody production. Fed. Proc. **24**, 179 (1965).

CAMPBELL, P. A., D. T. ROWLANDS, JR., M. J. HARRINGTON, and P. D. KIND: The adjuvant action of endotoxin in thymectomized mice. J. Immunol. **96**, 849—853 (1966).

CHERVENICK, P. A., D. R. BOGGS, J. C. MARSH, G. E. CARTWRIGHT, and M. M. WINTROBE: The blood and bone marrow neutrophil response to graded doses of endotoxin in mice. Proc. Soc. exp. Biol. (N.Y.) **126**, 891—895 (1967).

CLAMAN, H. N.: Tolerance to a protein antigen in adult mice and the effect of nonspecific factors. J. Immunol. **91**, 833—839 (1963).

— E. A. CHAPERON, and J. C. SELNER: Thymus-marrow immunocompetence. III. The requirement for living thymus cells. Proc. Soc. exp. Biol. (N.Y.) **127**, 462—466 (1968).

COHEN, E. P., L. K. CROSBY, and D. W. TALMAGE: The effect of endotoxin on the antibody response of transferred spleen cells. J. Immunol. **92**, 223—226 (1964).

COHN, Z. A.: The fate of bacteria within phagocytic cells. III. Destruction of an *Escherichia coli* agglutinogen within polymorphonuclear leucocytes and macrophages. J. exp. Med. **120**, 869—883 (1964).

COLLINS, R. D., and W. B. WOOD, JR.: Studies on the pathogenesis of fever. VI. The interaction of leucocytes and endotoxin *in vitro*. J. exp. Med. **110**, 1005—1016 (1959).

CONDIE, R. M., and R. A. GOOD: Inhibition of immunological enhancement by endotoxin in refractory rabbits. Immunochemical study. Proc. Soc. exp. Biol. (N.Y.) **91**, 414—418 (1956).

— S. J. ZAK, and R. A. GOOD: Effect of meningococcal endotoxin on resistance to bacterial infection and the immune response of rabbits. Fed. Proc. **14**, 459—460 (1955a).

— — — Effect of meningococcal endotoxin on the immune response. Proc. Soc. exp. Biol. (N.Y.) **90**, 355—360 (1955b).

DACHY, A.: De la sensibilité a l'action hémorragipare des endotoxines. Rev. Immunol. (Paris) **31**, 133—292 (1967).

DAVIES, A. M., I. GERY, E. ROSENMANN, and A. LAUFER: Endotoxin as adjuvant in autoimmunity to cardiac tissue. Proc. Soc. exp. Biol. (N.Y.) **114**, 520—523 (1963).

DIAZ, F., and E. NETER: Antibody response to the common enterobacterial antigen of children with shigellosis, salmonellosis or urinary tract infection. Amer. J. med. Sci. **256**, 18—24 (1968).

DOMINGUE, G. J., and E. NETER: Opsonizing and bactericidal activity of antibodies against common antigen of enterobacteriaceae. J. Bact. **91**, 129—133 (1966a).

— — Inhibition by lipopolysaccharide of immune phagocytosis of latex particles modified with common antigen of enteric bacteria. Proc. Soc. exp. Biol. (N.Y.) **121**, 133—137 (1966b).

— — The plaque test for the demonstration of antibodies against the enterobacterial common antigen produced by spleen and lymph node cells. Immunology **13**, 539—545 (1967).

DRESSER, D. W.: Effectiveness of lipid and lipidolphilic substances as adjuvants. Nature (Lond.) **191**, 1169—1171 (1961).
— Adjuvanticity of vitamin A. Nature (Lond.) **217**, 527—529 (1968).
DUBOS, R. J., and R. W. SCHAEDLER: Reversible changes in the susceptibility of mice to bacterial infections. I. Changes brought about by injection of pertussis vaccine or of bacterial endotoxins. J. exp. Med. **104**, 53—65 (1956).
DU BUY, B.: Role of the granulocyte in the pyrogenic response to intracisternal endotoxin. Proc. Soc. exp. Biol. (N.Y.) **123**, 606—609 (1966).
FARAGÓ, F., and S. PUSZTAI: An investigation into the effect of combined diphtheria-tetanus-pertussis prophylactic. Brit. J. exp. Path. **30**, 572—581 (1949).
FARTHING, J. R.: The role of *Bordetella pertussis* as an adjuvant to antibody production. Brit. J. exp. Path. **42**, 614—622 (1961).
—, and L. B. HOLT: Experiments designed to determine the mechanism of the adjuvant activity of gram-negative organisms upon antibody production. J. Hyg. (Lond.) **60**, 411—426 (1962).
FINGER, H.: Das Freundsche Adjuvans. Wesen und Bedeutung. Stuttgart: Gustav Fischer 1964.
— P. EMMERLING, and H. SCHMIDT: The influence of bacterial endotoxin on the formation of antibody-forming spleen cells in mice immunized with sheep red blood cells. Experientia (Basel) **23**, 849—850 (1967).
FINNEGAN, J. K., and J. B. DIENNA: Toxicological observations on certain surface-active agents. Proc. Sci. Sect. Toilet Goods Ass. **20**, 16—19 (1953).
FISHMAN, M., and F. L. ADLER: The role of macrophage-RNA in the immune response. Cold Spr. Harb. Symp. quant. Biol. **32**, 343—348 (1967).
FLEMING, D. S., L. GREENBERG, and E. M. BEITH: The use of combined antigens in the immunization of infants. Canad. med. Ass. J. **59**, 101—105 (1948).
FORBES, I. J.: Induction of mitosis in macrophages by endotoxin. J. Immunol. **94**, 37—39 (1965).
FOX, A. C., R. S. WILLIS, and H. H. FREEDMAN: Differential modification of the hemolysin response of mice to sheep erythrocytes and endotoxin. Bact. Proc. **74** (1968).
FREEDMAN, H. H., and W. BRAUN: Influence of oligodeoxyribonucleotides on phagocytic activity of the reticuloendothelial system. Proc. Soc. exp. Biol. (N.Y.) **120**, 222—225 (1965).
— A. E. FOX, and B. S. SCHWARTZ: Antibody formation at various times after previous treatment of mice with endotoxin. Proc. Soc. exp. Biol. (N.Y.) **125**, 583—587 (1967a).
— — R. S. WILLIS, and B. S. SCHWARTZ: Role of protein component of endotoxin in modification of host reactivity. Proc. Soc. exp. Biol. (N.Y.) **125**, 1316—1320 (1967b).
— — — — Induced sensitization of normal laboratory animals to *Brucella abortus* endotoxin. J. Bact. **95**, 286—290 (1968).
— M. NAKANO, and W. BRAUN: Antibody formation in endotoxin-tolerant mice. Proc. Soc. exp. Biol. (N.Y.) **121**, 1228—1230 (1966).
—, and B. M. SULTZER: Dissociation of the biological properties of bacterial endotoxin by chemical modification of the molecule. J. exp. Med. **116**, 929—942 (1962).
— — Aspects of endotoxin tolerance: Phagocytosis and specificity. In: Bacterial endotoxins. New Brunswick, N. J.: Rutgers University Press 1964.
FREI, P. C., B. BENACERRAF, and G. J. THORBECKE: Phagocytosis of the antigen, a crucial step in the induction of the primary response. Proc. nat. Acad. Sci. (Wash.) **53**, 20—23 (1965).
FRUHMAN, G. J.: Bacterial endotoxin: Effects on erythropoiesis. Blood **27**, 363—370 (1966).
GALLILY, R., and M. FELDMAN: The role of macrophages in the induction of antibody in x-irradiated animals. Immunology **12**, 197—206 (1967).

GANS, H., and W. KRIVIT: Effect of endotoxin on the clotting mechanism: II. On the variation in response in different species of animals. Ann. Surg. **153**, 453—458 (1961).

GARTNER, H., F. THIESSEN, R. HUNGERLAND u. H. P. FURST: Untersuchungen über die Wirkung von Endotoxinen an isolierten Organen. Arch. Hyg. (Berl.) **148**, 493—504 (1964).

GERSHON, R. K., V. WALLIS, A. J. S. DAVIES, and E. LEUCHARS: Inactivation of thymus cells after multiple injections of antigen. Nature (Lond.) **218**, 380—381 (1968).

GEWURZ, H., S. E. MERGENHAGEN, A. NOWOTNY, and J. K. PHILLIPS: Interactions of the complement system with native and chemically modified endotoxins. J. Bact. **95**, 397—405 (1968).

— H. S. SHIN, and S. E. MERGENHAGEN: Interactions of the complement system with endotoxic lipopolysaccharide: consumption of each of the six terminal complement components. J. exp. Med. **128**, 1049—1057 (1968).

GILBERT, R. P.: Endotoxin shock in the primate. Proc. Soc. exp. Biol. (N.Y.) **111**, 328—331 (1962).

GILBERT, V. E., and A. I. BRAUDE: Reduction of serum complement in rabbits after injection of endotoxin. J. exp. Med. **116**, 477—490 (1962).

GOOD, R. A., and A. E. GABRIELSEN (Eds.): The thymus in immunobiology. New York: Harper & Row 1964.

GORZYNSKI, E. A., and E. NETER: Immunization with common enterobacterial antigen against *Salmonella enteritidis*. Bact. Proc. 86 (1968).

— H. Y. WHANG, T. SUZUKI, and E. NETER: Differences in antibody response of rabbit to enterobacterial antigen after intravenous and subcutaneous injection with adjuvant. Proc. Soc. exp. Biol. (N.Y.) **114**, 700—704 (1963).

GREENBERG, L., and D. S. FLEMING: Increased efficiency of diphtheria toxoid when combined with pertussis vaccine. A preliminary note. Canad. J. publ. Hlth **38**, 279—282 (1947).

— — The immunizing efficiency of diphtheria toxoid when combined with various antigens. Canad. J. publ. Hlth **39**, 131—135 (1948).

GREISMAN, S. E., H. N. WAGNER, JR., M. IIO, and R. B. HORNICK: Mechanisms of endotoxin tolerance. II. Relationship between endotoxin tolerance and reticulo-endothelial system phagocytic activity in man. J. exp. Med. **119**, 241—265 (1964).

HARRIS, G.: Further studies of antigen stimulation of deoxyribonucleic acid synthesis in rabbit spleen cell cultures. II. The effects of specific antibody. Immunology **14**, 415—423 (1968).

—, and W.A. CRAMP: Further studies of antigen stimulation of deoxyribonucleic acid synthesis in rabbit spleen cell cultures. I. The effects of centrifuged proteins. Immunology **14**, 409—414 (1968).

HASKINS, W. T., M. LANDY, K. C. MILNER, and E. RIBI: Biological properties of parent endotoxins and lipoid fractions, with a kinetic study of acid-hydrolyzed endotoxin. J. exp. Med. **114**, 665—684 (1961).

HEILMAN, D. H.: The selective toxicity of endotoxin for phagocytic cells of the reticuloendothelial system. Int. Arch. Allergy **26**, 63—79 (1965).

— In vitro toxicity of endotoxin for macrophages of young guinea pigs and rabbits. Int. Arch. Allergy **33**, 501—510 (1968).

—, and R. BAST, JR.: Effect of endotoxin on macrophages of the adult chicken. Proc. Soc. exp. Biol. (N.Y.) **116**, 1019—1022 (1964).

— — *In vitro* assay of endotoxin by the inhibition of macrophage migration. J. Bact. **93**, 15—20 (1967).

HERION, J. C., R. I. WALKER, W. B. HERRING, and J. G. PALMER: Effects of endotoxin and nitrogen mustard on leukocyte kinetics. Blood **25**, 522—540 (1965).

HEUER, A. E., and B. PERNIS: Effect of endotoxin on the number of antibody-producing cells in mouse spleens. Bact. Proc. 44 (1964).

HILL, W. C., and D. ROWLEY: The origin of the antibody released into serum following injection of bacterial lipopolysaccharide. Aust. J. exp. Biol. med. Sci. **45**, 693—701 (1967).
HOLTON, J. B., and J. H. SCHWAB: Adjuvant properties of bacterial cell wall mucopeptides. J. Immunol. **96**, 134—138 (1966).
JAROSLOW, B. N.: Factors associated with initiation of the immune response. J. infect. Dis. **107**, 56—64 (1960).
JENKIN, C., and D. L. PALMER: Changes in the titre of serum opsonins and phagocytic properties of mouse peritoneal macrophages following injection of endotoxin. J. exp. Med. **112**, 419—429 (1960).
JOHNSON, A. G.: Adjuvant action of bacterial endotoxins on the primary antibody response. In: Bacterial endotoxins. New Brunswick, N. J.: Rutgers University Press 1964.
— S. GAINES, and M. LANDY: Studies on the O antigen of *Salmonella typhosa*. V. Enhancement of antibody response to protein antigens by the purified lipopolysaccharide. J. exp. Med. **103**, 225—246 (1956).
—, and A. NOWOTNY: Relationship of structure to function in bacterial O antigens. J. Bact. **87**, 809—814 (1964).
JOHNSON, H. G., and A. G. JOHNSON: Enhancement of antibody synthesis in mice by macrophages stimulated *in vitro* with antigen and poly A-U. Bact. Proc. 75—76 (1968).
KAMRIN, B. B.: Alpha globulin injections and decreased gamma globulin production in chickens. Science **153**, 1261—1262 (1966).
KASAI, N., and A. NOWOTNY: Endotoxic glycolipid from a heptoseless mutant of *Salmonella minnesota*. J. Bact. **94**, 1824—1836 (1967).
KASS, E. H., R. P. ATWOOD, and P. J. PORTER: Observations on the locus of lethal action of bacterial endotoxin. In: Bacterial endotoxins. New Brunswick, N.J.: Rutgers University Press 1964.
KESSEL, R. W. I., and W. BRAUN: Cytotoxicity of endotoxin *in vitro*. Effects on macrophages from normal guinea pigs. Aust. J. exp. Biol. med. Sci. **43**, 511—522 (1965).
— —, and O. J. PLESCIA: Possible role of cell-associated antibody in the cytotoxicity of endotoxin for macrophages. Fed. Proc. **23**, 564 (1964).
— — — Endotoxin cytotoxicity: Role of cell-associated antibody. Proc. Soc. exp. Biol. (N.Y.) **121**, 449—452 (1966a).
— H. H. FREEDMAN, and W. BRAUN: Relation of polysaccharide content to some biological properties of endotoxins from mutants of *Salmonella typhimurium*. J. Bact. **92**, 592—596 (1966b).
— M. HECHTEL, and W. BRAUN: Endotoxin activity in a coliphage preparation. J. Bact. **91**, 1382—1383 (1966c).
— E. NETER, and W. BRAUN: Biological activities of the common antigen of enterobacteriaceae. J. Bact. **91**, 465—466 (1966d).
KHANOLKAR, V. R.: Non-specific training of antibody production. J. Path. **27**, 181—186 (1924).
KIERSZENBAUM, F., N. R. NOTA, R. B. MARRA, and S. C. BESUSCHIO: Cell participation in the stimulation of immunologic response by *Escherichia coli* lipopolysaccharide. Int. Arch. Allergy **32**, 91—97 (1967).
KIM, Y. B., and D. W. WATSON: Inactivation of gram-negative bacterial endotoxins by papain. Proc. Soc. exp. Biol. (N.Y.) **115**, 140—142 (1964).
— — Modification of host responses to bacterial endotoxins. II. Passive transfer of immunity to bacterial endotoxin with fractions containing 19S antibodies. J. exp. Med. **121**, 751—759 (1965).
— — Role of antibodies in reactions to gram-negative bacterial endotoxins. Ann. N.Y. Acad. Sci. **133**, 727—745 (1966).

KIM, Y. B., and D. W. WATSON: Biologically active endotoxins from *Salmonella* mutants deficient in O- and R-polysaccharides and heptose. J. Bact. **94**, 1320—1326 (1967).
KIMBALL, H. R., T. W. WILLIAMS, and S. M. WOLFF: Effect of bacterial endotoxin on experimental fungal infections. J. Immunol. **100**, 24—33 (1968).
KIND, P., P. CAMPBELL, and D. T. ROWLANDS, JR.: Endotoxin-induced wasting disease in mice: A temporary condition explained by endotoxin tolerance. Proc. Soc. exp. Biol. (N.Y.) **125**, 495—498 (1967).
—, and A. G. JOHNSON: Studies on the adjuvant action of bacterial endotoxins on antibody formation. I. Time limitation of enhancing effect and restoration of antibody formation in x-irradiated rabbits. J. Immunol. **82**, 415—427 (1959).
KONG, Y.-C. M., and A. G. JOHNSON: Factors affecting primary and secondary antibody production by splenic tissues. J. Immunol. **90**, 672—684 (1963).
KUNIN, G. M.: Separation, characterization, and biological significance of a common antigen in enterobacteriaceae. J. exp. Med. **118**, 565—586 (1963).
— M. V. BEARD, and N. E. HALMAGYI: Evidence for a common hapten associated with endotoxin fractions of *E. coli* and other enterobacteriaceae. Proc. Soc. exp. Biol. (N.Y.) **111**, 160—166 (1962).
LANDY, M.: Increase in resistance following administration of bacterial lipopolysaccharides. Ann. N.Y. Acad. Sci. **66**, 292—303 (1956).
— R. P. SANDERSON, M. T. BERNSTEIN, and E. M. LERNER, II: Involvement of thymus in immune response of rabbits to somatic polysaccharides of gram-negative bacteria. Science **147**, 1591—1592 (1965).
LANGEVOORT, H. L., R. M. ASOFSKY, E. B. JACOBSON, T. DE VRIES, and G. J. THORBECKE: Gamma-globulin and antibody formation *in vitro*. II. Parallel observations on histologic changes and on antibody formation in the white and red pulp of the rabbit spleen during the primary response, with special reference to the effect of endotoxin. J. Immunol. **90**, 60—71 (1963).
LEVIN, J.: Blood coagulation and endotoxin in invertebrates. Fed. Proc. **26**, 1707—1712 (1967).
LEWIS, P. A., and D. LOOMIS: Allergic irritability. III. The influence of chronic infections and of trypan blue on the formation of specific antibodies. J. exp. Med. **43**, 263—273 (1926).
LUECKE, D. H., and L. R. SIBAL: Enhancement by endotoxin of the primary antibody response to bovine serum albumin in chickens. J. Immunol. **89**, 539—544 (1962).
LÜDERITZ, O., K. JANN, and R. WHEAT: Somatic and capsular antigens of gram-negative bacteria. In: Comprehensive biochemistry, vol. 26A. Amsterdam: Elsevier Publ. Co. 1968.
— A. M. STAUB, and O. WESTPHAL: Immunochemistry of O and R antigens of *Salmonella* and related enterobacteriaceae. Bact. Rev. **30**, 192—255 (1966).
— O. WESTPHAL, E. EICHENBERGER u. E. NETER: Über die Komplexbildung von bakteriellen Lipopolysacchariden mit Proteinen und Lipoiden. Biochem. Z. **330**, 21—33 (1958).
MACLEAN, I. H., M. B. EDIN, and L. B. HOLT: Combined immunization with tetanus toxoid and T.A.B. Lancet **1940** II, 581—583.
MARGHERITA, S. S., and H. FRIEDMAN: Induction of nonspecific resistance by endotoxin in unresponsive mice. J. Bact. **89**, 277—280 (1965).
—, and R. A. PATNODE: Modification of an endotoxin effect by esters. Experientia (Basel) **24**, 267—268 (1968).
MARTIN, W. J., and S. MARCUS: Detoxified bacterial endotoxins. I. Preparation and biological properties of an acetylated crude endotoxin from *Salmonella typhimurium*. J. Bact. **91**, 1453—1459 (1966a).
— — Detoxified bacterial endotoxins. II. Preparation and biological properties of chemically modified crude endotoxins from *Salmonella typhimurium*. J. Bact. **91**, 1750—1758 (1966b).

McKay, D. G., and T.-C. Wong: The effect of bacterial endotoxin on the placenta of the rat. Amer. J. Path. **42**, 357—377 (1963).

McKenna, J. M., and K. M. Stevens: The early phase of the antibody response. J. Immunol. **78**, 311—317 (1957).

Mergenhagen, S. E., H. A. Bladen, and K. C. Hsu: Electron microscopic localization of endotoxic lipopolysaccharide in gram-negative organisms. Ann. N.Y. Acad. Sci. **133**, 279—291 (1966).

— H. Gewurz, H. A. Bladen, A. Nowotny, N. Kasai, and O. Lüderitz: Interactions of the complement system with endotoxins from a *Salmonella minnesota* mutant deficient in O-polysaccharide and heptose. J. Immunol. **100**, 227—229 (1968).

—, and S. B. Jensen: Specific induced resistance of mice to endotoxin. Proc. Soc. exp. Biol. (N.Y.) **110**, 139—142 (1962).

— A. L. Notkins, and S. F. Dougherty: Adjuvanticity of lactic dehydrogenase virus: Influence of virus infection on the establishment of immunologic tolerance to a protein antigen in adult mice. J. Immunol. **99**, 576—581 (1967).

— E. Schiffmann, G. R. Martin, and S. B. Jensen: Stimulation of resistance to infection by lipoid derived from *Neisseria* meningitidis endotoxin. J. infect. Dis. **113**, 52—58 (1963).

Merritt, K., and A. G. Johnson: Studies on the adjuvant action of bacterial endotoxins on antibody formation. V. The influence of endotoxin and 5-fluoro-2-deoxyuridine on the primary antibody response of the Balb mouse to a purified protein antigen. J. Immunol. **91**, 266—272 (1963).

— — Studies on the adjuvant action of bacterial endotoxins on antibody formation. VI. Enhancement of antibody formation by nucleic acids. J. Immunol. **94**, 416—422 (1965).

Metcalf, D.: Relation of the thymus to the formation of immunologically reactive cells. Cold Spr. Harb. Symp. quant. Biol. **32**, 583—590 (1967).

Michael, J. G.: The release of specific bactericidal antibodies by endotoxin. J. exp. Med. **123**, 205—212 (1966).

— J. L. Whitby, and M. Landy: Increase in specific bactericidal antibodies after administration of endotoxin. Nature (Lond.) **191**, 296—297 (1961).

Mihich, E., O. Westphal, O. Lüderitz, and E. Neter: The tumor necrotizing effect of lipoid A component of *Escherichia coli* endotoxin. Proc. Soc. exp. Biol. (N.Y.) **107**, 816—819 (1961).

Miler, I.: The effect of bacterial endotoxin on the phagocytic activity of the reticulo-endothelial system in rabbits of different age. Folia microbiol. (Praha) **8**, 265—273 (1963).

Milner, K. C., and R. A. Finkelstein: Bioassay of endotoxin: Correlation between pyrogenicity for rabbits and lethality for chick embryos. J. infect. Dis. **116**, 529—536 (1966).

—, and J. A. Rudbach: Relationship between presence of specific antibody and tolerance to endotoxin. Bact. Proc. 96 (1968).

Mosier, D. E.: A requirement for two cell types for antibody formation *in vitro*. Science **158**, 1573—1575 (1967).

Neeper, C. A., and C. V. Seastone: Mechanisms of immunologic paralysis by pneumococcal polysaccharide. II. The influence of nonspecific factors on the immunity of paralyzed mice to pneumococcal infection. J. Immunol. **91**, 378—383 (1963).

Neter, E.: Bacterial hemagglutination and hemolysis. Bact. Rev. **20**, 166—188 (1956).

— Indirect bacterial hemagglutination, and its application to the study of bacterial antigens and serologic diagnosis. Path. et Microbiol. (Basel) **28**, 859—877 (1965).

NETER, E., H. ANZAI, E. A. GORZYNSKI, A. NOWOTNY, and O. WESTPHAL: Effects of lipoid A component of *Escherichia coli* endotoxin on dermal reactivity of rabbits to epinephrine. Proc. Soc. exp. Biol. (N.Y.) **103**, 783—786 (1960).
— E. A. GORZYNSKI, O. WESTPHAL, and O. LÜDERITZ: The effects of antibiotics on enterobacterial lipopolysaccharides (endotoxins), hemagglutination and hemolysis. J. Immunol. **80**, 66—72 (1958a).
— — — —, and D. J. KLUMPP: The effects of protamine and histone on enterobacterial lipopolysaccharides and hemolysis. Canad. J. Microbiol. **4**, 371—383 (1958b).
— O. WESTPHAL, and O. LÜDERITZ: Effects of lecithin, cholesterol, and serum on erythrocyte modification and antibody neutralization by enterobacterial lipopolysaccharides. Proc. Soc. exp. Biol. (N.Y.) **88**, 339—341 (1955).
— — —, and E. A. GORZYNSKI: The bacterial hemagglutination test for the demonstration of antibodies to enterobacteriaceae. Ann. N.Y. Acad. Sci. **66**, 141—156 (1956).
— H. Y. WHANG, O. LÜDERITZ, and O. WESTPHAL: Immunological priming without production of circulating bacterial antibodies conditioned by endotoxin and its lipid A component. Nature (Lond.) **212**, 420—421 (1966).
— — T. SUZUKI, and E. A. GORZYNSKI: Differences in antibody response of rabbit to intravenously injected soluble and cell-attached enterobacterial antigen. Immunology **7**, 657—664 (1964).
— N. J. ZALEWSKI, and D. A. ZAK: Inhibition by lecithin and cholesterol of bacterial (*Escherichia coli*) hemagglutination and hemolysis. J. Immunol. **71**, 145—151 (1953).
NICOL, T., D. C. QUANTOCK, and B. VERNON-ROBERTS: Stimulation of phagocytosis in relation to the mechanism of action of adjuvants. Nature (Lond.) **209**, 1142—1143 (1966).
NOLAN, J. P., and C. J. O'CONNELL: Vascular response in the isolated rat liver. J. exp. Med. **122**, 1063—1073 (1965).
NOWOTNY, A.: Endotoxoid preparations. Nature (Lond.) **197**, 721—722 (1963).
— Chemical detoxification of bacterial endotoxins. In: Bacterial endotoxins. New Brunswick, N. J.: Rutgers University Press 1964.
— Immunogenicity of toxic and detoxified endotoxin preparations. Proc. Soc. exp. Biol. (N.Y.) **127**, 745—748 (1968).
— R. RADVANY, and N. E. NEALE: Neutralization of toxic bacterial O-antigens with O-antibodies while maintaining their stimulus on non-specific resistance. Life Sci. **4**, 1107—1114 (1965).
OH, J. O.: An interferon-like viral inhibitor in body fluids of endotoxin-injected rabbits. Proc. Soc. exp. Biol. (N.Y.) **123**, 493—496 (1966).
ØRSKOV, F., I. ØRSKOV, B. JANN, K. JANN, E. MÜLLER-SEITZ, and O. WESTPHAL: Immunochemistry of *Escherichia coli* O antigens. Acta path. microbiol. scand. **71**, 339—358 (1967).
OSOBA, D.: Thymic control of cellular differentiation in the immunological system. Proc. Soc. exp. Biol. (N.Y.) **127**, 418—420 (1968).
PATTERSON, L. T., J. M. HARPER, and R. D. HIGGINBOTHAM: Association of C-reactive protein and circulating leukocytes with resistance to *Staphylococcus aureus* infection in endotoxin-treated mice and rabbits. J. Bact. **95**, 1375—1379 (1968).
—, and R. D. HIGGINBOTHAM: Mouse C-reactive protein and endotoxin-induced resistance. J. Bact. **90**, 1520—1524 (1965).
PAVLOVSKIS, O., and G. F. SPRINGER: Lipopolysaccharide receptor from human erythrocytes. Fed. Proc. **26**, 801 (1967).
PEARLMAN, D. S., J. B. SAUERS, and D. W. TALMAGE: The effect of adjuvant amounts of endotoxins on the serum hemolytic complement activity in rabbits. J. Immunol. **91**, 748—756 (1963).

PETERSON, K., and U. PAVEL: Effect of endotoxin on resistance in the early period of ontogenesis. Nature (Lond.) **214**, 699—700 (1967).

PIERCE, C. W.: The effects of endotoxin on the immune response in the rat. I. Antibodies formed against bovine gamma-globulin in the intact rat. Lab. Invest. **16**, 768—781 (1967a).

— The effects of endotoxin on the immune response in the rat. II. Antibodies formed against bovine gamma-globulin in the splenectomized rat. Lab. Invest. **16**, 782—794 (1967b).

— The effects of endotoxin on the immune response in the rat. III. Elimination of I^{125}-labeled bovine gamma-globulin from the circulation of rats. Lab. Invest. **17**, 380—392 (1967c).

PIERONI, R. E., and L. LEVINE: Mouse strains selectivity in the endotoxin-induced hypersensitivity to histamine. J. Bact. **95**, 716—717 (1968).

PINCHUCK, P., M. FISHMAN, F. L. ADLER, and P. H. MAURER: Antibody formation: Initiation in "Nonresponder" mice by macrophage synthetic polypeptide RNA. Science **160**, 194—195 (1968).

PINCKARD, R. N., D. M. WEIR, and W. H. MCBRIDE: Effects of *Corynebacterium parvum* on immunological unresponsiveness to bovine serum albumin in the rabbit. Nature (Lond.) **215**, 870—871 (1967).

PLESCIA, O. J., and W. BRAUN: Nucleic acids as antigens. In: Advances in immunology. New York: Academic Press 1967.

— — (Eds.): Nucleic acids in immunology. Berlin-Heidelberg-New York: Springer 1968.

PREVITE, J. J., Y. CHANG, and H. M. EL-BISI: Detoxification of *Salmonella typhimurium* lipopolysaccharide by ionizing radiation. J. Bact. **93**, 1607—1614 (1967).

PROCHAZKA, O.: Stimulation of antibody formation by bacterial lipopolysaccharide. The influence of repeated doses of lipopolysaccharide on antibody formation. Folia microbiol. (Praha) **6**, 157—163 (1961).

RADVANY, R., N. L. NEALE, and A. NOWOTNY: Relation of structure to function in bacterial O-antigens. VI. Neutralization of endotoxic O-antigens by homologous O-antibody. Ann. N.Y. Acad. Sci. **133**, 763—786 (1966).

RAMON, G., et C. ZOELLER: Les "vaccins associés" par union d'une anatoxine et d'un vaccin microbien (TAB) ou par mélanges d'anatoxines. C. R. Soc. Biol. (Paris) **94**, 106—109 (1926).

RANTZ, L. A., E. RANDALL, and A. ZUCKERMAN: Hemolysis and hemagglutination by normal and immune serums of erythrocytes treated with a nonspecies specific bacterial substance. J. infect. Dis. **98**, 211—222 (1956).

RAUSS, K., I. KÉTYI u. L. RÉTHY: Untersuchungen über die Entgiftung und Adjuvierung von enteralen Impfstoffen. Z. Immun.-Forsch. **116**, 276—286 (1958).

REGAMEY, R. H., W. HENNESSEN, D. IKIĆ, and J. UNGAR (eds.): International symposium on adjuvants of immunity. Basel: S. Karger 1967.

RÉTHY, L.: Further investigations on the adjuvanting action of endotoxins. Ann. Immunol. Hung. **10**, 127—131 (1967).

ROSENBLATT, E., and A. G. JOHNSON: Size of antibody and role of autonomic nervous system in the adjuvant action of endotoxin. Proc. Soc. exp. Biol. (N.Y.) **113**, 156—161 (1963).

ROSSE, W. F., T. BORSOS, and H. J. RAPP: Cold-reacting antibodies: The enhancement of antibody fixation by the first component of complement (C'la). J. Immunol. **100**, 259—265 (1968).

ROTHFIELD, L., M. TAKESHITA, M. PEARLMAN, and R. W. HORNE: Role of phospholipids in the enzymatic synthesis of the bacterial cell envelope. Fed. Proc. **25**, 1495—1502 (1966).

ROWLANDS, D. T., JR., H. N. CLAMAN, and P. D. KIND: The effect of endotoxin on the thymus of young mice. Amer. J. Path. **46**, 165—176 (1965).
ROWLEY, D.: Rapidly induced changes in the level of non-specific immunity in laboratory animals. Brit. J. exp. Path. **37**, 223—234 (1956).
— The role of opsonins in non-specific immunity. J. exp. Med. **111**, 137—144 (1960).
—, and K. J. TURNER: Increase in macroglobulin antibodies of mouse and pig following injection of bacterial lipopolysaccharide. Immunology **7**, 394—402 (1964).
RUDBACH, J. A., K. C. MILNER, and E. RIBI: Hybrid formation between bacterial endotoxins. J. exp. Med. **126**, 63—79 (1967).
SASSEN, A., E. H. PERKINS, and R. A. BROWN: Immunogenic potency of human gamma-globulin in mice. Immunology **14**, 247—256 (1968).
SEIBERT, F.: Fever-producing substance found in some distilled waters. Amer. J. Physiol. **67**, 90—104 (1923).
—, and L. B. MENDEL: Protein fevers. With special reference to casein. Amer. J. Physiol. **67**, 105—123 (1923).
SHANDS, J. W., JR., J. A. GRAHAM, and K. NATH: The morphologic structure of isolated bacterial lipopolysaccharide. J. molec. Biol. **25**, 15—21 (1967).
SHUMWAY, C. N., V. BOKKENHEUSER, D. POLLOCK, and E. NETER: Survival in immune and nonimmune rabbits of Cr^{51}-labeled erythrocytes modified by bacterial antigen. J. Lab. clin. Med. **62**, 600—607 (1963).
SIBAL, L. R.: Effect of endotoxin on antibody production by chicken spleen cells transferred to chick chorioallantois. J. Immunol. **87**, 362—366 (1961).
— M. A. FINK, D. D. ROBERTSON, and C. A. COWLES: Enhanced production of antibody to a murine leukemia virus (Rauscher) in the host of origin. Proc. Soc. exp. Biol. (N.Y.) **127**, 726—730 (1968).
SINGER, I., E. T. KIMBLE, III, and R. E. RITTS, JR.: Alterations of the host-parasite relationship by administration of endotoxin to mice with infections of trypanosomes. J. infect. Dis. **114**, 243—248 (1964).
SNYDERMAN, R., H. GEWURZ, and S. E. MERGENHAGEN: Generation of factors chemotactic for polymorphonuclear leukocytes (PMN) by endotoxin (LPS) and complement (C′). Fed. Proc. **27**, 355 (1968a).
— — — Interactions of the complement system with endotoxic lipopolysaccharide: Generation of a factor chemotactic for polymorphonuclear leukocytes. J. exp. Med. **128**, 259—275, (1968b).
SOLLIDAY, S., D. A. ROWLEY, and F. W. FITCH: Adjuvant reversal of antibody suppression of the immune response. Fed. Proc. **26**, 700 (1967).
SPIELVOGEL, A. R.: An ultrastructural study of the mechanisms of platelet-endotoxin interaction. J. exp. Med. **126**, 235—250 (1967).
SPINK, W. W., and J. VICK: A labile serum factor in experimental endotoxin shock: cross-transfusion studies in dogs. J. exp. Med. **114**, 501—508 (1961).
SPRINGER, G. F., and R. E. HORTON: Erythrocyte sensitization by blood group-specific bacterial antigens. J. gen. Physiol. **47**, 1229—1250 (1964).
— E. T. WANG, J. H. NICHOLS, and J. M. SHEAR: Relations between bacterial lipopolysaccharide structures and those of human cells. Ann. N.Y. Acad. Sci. **133**, 566—579 (1966).
SPURGASH, A. J., and L. R. SIBAL: Formation of antibody to a protein antigen in adult rabbits given *Salmonella lipopolysaccharide*. Proc. Soc. exp. Biol. (N.Y.) **125**, 1112—1115 (1967).
STERZL, J., M. HOLUB, and I. MILER: Effect of endotoxin on antibody response and resistance to infection in newborn animals. Folia microbiol. (Praha) **6**, 289—298 (1961).
STETSON, C. A.: Role of hypersensitivity in reactions to endotoxin. In: Bacterial endotoxins. New Brunswick, N. J.: Rutgers University Press 1964.

STEVENS, K. M., and J. M. MCKENNA: Studies on antibody synthesis initiated in vitro. J. exp. Med. **107**, 537—559 (1958).

STINEBRING, W. R., and J. S. YOUNGNER: Patterns of interferon appearance in mice injected with bacteria or bacterial endotoxin. Nature (Lond.) **204**, 712 (1964).

STUART, A. E., and A. E. DAVIDSON: Effect of simple lipids on antibody formation after injection of foreign red cells. J. Path. Bact. **87**, 305—316 (1964).

SUZUKI, T., E. A. GORZYNSKI, and E. NETER: Separation by ethanol of common and somatic antigens of enterobacteriaceae. J. Bact. **88**, 1240—1243 (1964a).

— H. Y. WHANG, E. A. GORZYNSKI, and E. NETER: Inhibition by lipopolysaccharide (endotoxin) of antibody response of rabbit to common antigen of enterobacteriaceae. Proc. Soc. exp. Biol. (N.Y.) **117**, 785—789 (1964b).

— —, and E. NETER: Studies of common antigen of enterobacteriaceae with particular reference to *Escherichia coli* O14. Ann. Immunol. Hung. **9**, 283—292 (1966).

TALIAFERRO, W. H., and B. N. JAROSLOW: The restoration of hemolysin formation in x-rayed rabbits by nucleic acid derivatives and antagonists of nucleic acid synthesis. J. infect. Dis. **107**, 341—350 (1960).

TALMAGE, D. W., and D. S. PEARLMAN: The antibody response: A model based on antagonistic actions of antigen. J. theor. Biol. **5**, 321—339 (1963).

TARMINA, D. F., K. C. MILNER, E. RIBI, and J. A. RUDBACH: Reaction of endotoxin and surfactants. II. Immunologic properties of endotoxins treated with sodium deoxycholate. J. Immunol. **100**, 444—450 (1968).

THORBECKE, G. J., R. M. ASOFSKY, G. M. HOCHWALD, and G. W. SISKIND: Gamma globulin and antibody formation *in vitro*. III. Induction of secondary response at different intervals after the primary; the role of secondary nodules in the preparation for the secondary response. J. exp. Med. **116**, 295—309 (1962).

TRENTIN, J. J. (ed.): Cross-reacting antigens and neoantigens. Baltimore: Williams & Wilkins Co. 1967.

TUROWSKA, B., and G. TUROWSKI: The endotoxin effect on the Gc (Group specific system) antiserum production. Z. Immun.-Forsch. Allerg. **129**, 368—373 (1965).

UCHITEL, I. Y., and E. L. KHASMAN: On the mechanism of adjuvant action of nonspecific stimulators of antibody formation. J. Immunol. **94**, 492—497 (1965).

UNANUE, E. R., and B. A. ASKONAS: Persistence of immunogenicity of antigen after uptake by macrophages. J. exp. Med. **127**, 915—926 (1968).

URBASCHEK, B., and A. NOWOTNY: Endotoxin tolerance induced by detoxified endotoxin (endotoxoid). Proc. Soc. exp. Biol. (N.Y.) **127**, 650—652 (1968).

WAGNER, R. R., R. M. SNYDER, E. W. HOOK, and C. N. LUTTRELL: Effect of bacterial endotoxin on resistance of mice to viral encephalitides. Including comparative studies of the interference phenomenon. J. Immunol. **83**, 87—98 (1959).

WARD, P. A., M. R. ABELL, and A. G. JOHNSON: Studies on the adjuvant action of bacterial endotoxins on antibody formation. IV. Histologic study of cortisone-treated rabbits. Amer. J. Path. **38**, 189—205 (1961).

—, and A. G. JOHNSON: Studies on the adjuvant action of bacterial endotoxins on antibody formation. II. Antibody formation in cortisone-treated rabbits. J. Immunol. **82**, 428—434 (1959).

— —, and M. R. ABELL: Studies on the adjuvant action of bacterial endotoxins on antibody formation. III. Histologic response of the rabbit spleen to a single injection of a purified protein antigen. J. exp. Med. **109**, 463—474 (1959).

WATSON, D. W., and Y. B. KIM: Modification of host responses to bacterial endotoxins. I. Specificity of pyrogenic tolerance and the role of hypersensitivity in pyrogenicity, lethality, and skin reactivity. J. exp. Med. **118**, 425—446 (1963).

— — Immunological aspects of pyrogenic tolerance. In: Bacterial endotoxins. New Brunswick, N. J.: Rutgers University Press 1964.

WEBB, P. M., and L. H. MUSCHEL: Bactericidal antibody studies in maternal serum, colostrum, and newborn serum of the pig. Canad. J. Microbiol. **14**, 545—550 (1968).

WEINER, R., and B. W. ZWEIFACH: Influence of *E. coli* endotoxin on serotonin contractions of the rabbit aortic strip. Proc. Soc. exp. Biol. (N.Y.) **123**, 937—939 (1966).

WEISSMANN, G.: Labilization and stabilization of lysosomes. Fed. Proc. **23**, 1038—1044 (1964).

WESTPHAL, O.: Pyrogens. In: Polysaccharides in biology. Madison, N. J.: Madison Printing Co. 1967.

— A. NOWOTNY, O. LÜDERITZ, H. HURNI, E. EICHENBERGER u. G. SCHONHOLZER: Die Bedeutung der Lipoid-Komponente (Lipoid A) für die biologischen Wirkungen bakterieller Endotoxine (lipopolysaccharide). Pharm. Acta Helv. **33**, 401—411 (1958).

WHANG, H. Y., O. LÜDERITZ, O. WESTPHAL, and E. NETER: Inhibition by lipoid A of formation of antibodies against common antigen of enterobacteriaceae. Proc. Soc. exp. Biol. (N.Y.) **120**, 371—374 (1965).

—, and E. NETER: Immunological studies of a heterogenetic enterobacterial antigen (Kunin). J. Bact. **84**, 1245—1250 (1962).

— — Study of heterogenetic (Kunin) antibodies in serum of healthy subjects and children with enteric and urinary tract infections. J. Pediat. **63**, 412—419 (1963).

— — Selective destruction by *Pseudomonas aeruginosa* of common antigen of enterobacteriaceae. J. Bact. **88**, 1244—1248 (1964).

— — Demonstration of common antigen in *Salmonella typhi* spheroplasts. Canad. J. Microbiol. **11**, 598—601 (1965a).

— — Destruction of common antigen of enterobacteriaceae by a psychrophilic pseudomonas. J. Bact. **89**, 1436—1437 (1965b).

— — Immunosuppression by endotoxin and its lipoid A component. Proc. Soc. exp. Biol. (N.Y.) **124**, 919—924 (1967a).

— — Further studies on effect of endotoxin on antibody response of rabbit to common antigen of enterobacteriaceae. J. Immunol. **98**, 948—957 (1967b).

— — Inhibition by cardiolipin of the antibody response to bacterial antigens. J. Immunol. **100**, 501—506 (1968a).

— — Effect of cholesterol on immunogenicity of common enterobacterial antigens. Proc. Soc. exp. Biol. (N.Y.) **128**, 956—959 (1968b).

— Y. YAGI, and E. NETER: Characterization of rabbit antibodies against common bacterial antigens and their presence in the fetus. Int. Arch. Allergy **32**, 353—365 (1967).

WHITBY, J. L., J. G. MICHAEL, M. W. WOODS, and M. LANDY: Symposium on bacterial endotoxins. II. Possible mechanisms whereby endotoxins evoke increased nonspecific resistance to infection. Bact. Rev. **25**, 437—446 (1961).

WIENER, R., A. BECK, and M. SHILO: Effect of bacterial lipopolysaccharides on mouse peritoneal leukocytes. Lab. Invest. **14**, 475—487 (1965).

WOLFF, S. M., J. H. MULHOLLAND, S. B. WARD, M. RUBENSTEIN, and P. D. MOTT: Effect of 6-mercaptopurine on endotoxin tolerance. J. clin. Invest. **44**, 1402—1409 (1965).

YOUNGNER, J. S., and W. R. STINEBRING: Comparison of interferon production in mice by bacterial endotoxin and statolon. Virology **29**, 310—316 (1966).

Institute of Virology, Czechoslovak Academy of Sciences, Bratislava

Serum Inhibitors of Myxoviruses

OLGA KRIŽANOVÁ, and VOJTA RATHOVÁ

With 5 Figures

Table of Contents

The term "serum inhibitors of myxoviruses" has been used to designate proteinaceous substances differing from antibody and occurring in animal sera. These substances prevent agglutination of red blood cells by myxoviruses and at least some of them also inhibit reproduction of myxoviruses in susceptible cells.

To distinguish them from specific antibody, the term "non-specific inhibitors" has been employed in the literature. This term is not fully adequate since some of the inhibitors are to a certain degree "specific", namely strain-specific (they inhibit only certain viral strains).

The terms alpha-, beta-, and gamma-inhibitor have been introduced to designate different types of serum inhibitors of influenza viruses. These designations, corresponding to the initial characters of Greek alphabet, have been given to inhibitors consecutively after it became clear that not a single component of serum, but substances of a different nature and possessing different properties were involved.

The terms "Francis' inhibitor" (equivalent to alpha-inhibitor) and "Chu's inhibitor" (equivalent to beta-inhibitor), derived from the discoverer's name (FRANCIS, 1947; CHU, 1951), are also frequent. In the second case we think that the name of MCCREA has been disregarded, since MCCREA already in 1946 described a thermolabile inhibitor in some animal sera, the properties of which correspond to those of beta-inhibitor.

With gamma-inhibitor, a designation according to the discoverer's name has not been used. This inhibitor was first named by SHIMOJO (1958) to distinguish it from alpha- and beta-inhibitor (before SHIMOJO, several authors used the name "A2 influenza virus inhibitor" to emphasize its considerable

Table 1. *Characteristics of influenza virus inhibitors*

	Inhibitor type		
	α	β	γ
Virus types inhibited	A, A1, A2, B	A, A1	A2
Neutralization of infectivity	—	+	+
Difference between inhibition of live and indicator virus	+++	—	± or +
Difference between inhibition of mouse-adapted and non-adapted virus	—	+++	± or +
Effect of heat	Unaffected or slightly increased titre	inactivated	increased titre
Effect of CO_2	Partially removed	partially removed	unaffected
Inactivation by:			
neuraminidase	+	—	—
trypsin	+	+	±
periodate	+	—	+
citrate	—	+	—
Distribution	Sera of most species. Some mucoids	bovine, mouse, rabbit, ferret, guinea-pig	horse, guinea-pig, rabbit, ferret, swine, human

strain-specificity). A designation according to the virus strain against which the inhibitor is active has been proposed by Styk (1955) for the C-inhibitor (i.e. C influenza virus inhibitor).

In recent years data on serum inhibitors have been accumulating so fast that the information started to become confusing.

Several authors tried to summarize and classify the knowledge about non-specific inhibition of viruses in the form of reviews (e.g. Jandásek, 1964; Niculescu, 1965; Orsi, 1966; Borecký, 1967).

We have included in the present report material published up to December, 1967. But we have to point out that in spite of our efforts, some data were not accessible in the original.

To specify the subject of our report, we are presenting the characteristics of some of the best known serum inhibitors in Table 1. It has been compiled mainly from data by Shimojo (1958), Cohen et al. (1963a), Orsi et al. (1967).

It is evident that the various inhibitors show different specific activities against different strains of viruses, different modes of interaction (binding) with the virus, a distinct susceptibility to the action of certain chemicals and to physical effects, and specific differences in the occurrence in sera of different animals.

Considering the ability of non-specific inhibitors to modify various manifestations of the viral activity, our discussion will be divided into three parts. We shall deal with:

1) the nature of non-specific inhibitors and their interaction with influenza virus;

2) the methods of removal of non-specific inhibitors from immune sera used either in studies on the antigenic structure of viruses, or for routine diagnostic purposes; and

3) the problem of the possible role of non-specific inhibitors in the natural resistance of animal organisms against viral diseases.

The Nature of Non-Specific Inhibitors

Based on results published hitherto, two groups of serum inhibitors may be distinguished: the inhibitors of glycoprotein nature, to which belong the alpha-, gamma-, and C inhibitors, and those of protein nature, like the beta-inhibitor. Because of several specific features of the individual inhibitors it seems useful to deal with each of them separately.

Alpha-Inhibitor

Among the serum inhibitors, the alpha-inhibitor resembles most closely the red blood cell receptors of influenza virus (Bowarnick et al., 1947; Gottschalk, 1959; Kathan et al., 1959, 1961, 1963) and eventually also receptors of cells supporting virus multiplication (Križanová-Laučíková et al., 1961a; Niculescu, 1965; Lobodzińska et al., 1966). Both alpha-inhibitors and receptors are of glycoprotein nature; they both inhibit only influenza virus haemagglutination and have no virus-neutralizing activity. The active group in both of them is sialic acid. The alpha-inhibitor is thermostable at 56° C for 30—60 minutes and inhibits haemagglutination by strains of A and B influenza virus (Smith et al., 1949). This inhibitor reacts more effectively with "indicator" virus, i.e. virus stripped of sialidase activity by heating at 56° C for 30 minutes (Stone, 1949), than with native (unheated) virus. Receptor-destroying enzyme (RDE) as well as the native (Isaacs et al., 1951; Borecký et al., 1962a) or ether-disrupted (Schmidt et al., 1960) virus split off sialic acid from alpha-inhibitor. These reactions result in a loss of its haemagglutination-inhibiting activity.

Nowadays it is generally accepted that it is not solely the presence of sialic acid but presumably its specific steric arrangement in the molecule of the inhibitor, which seems to be important for the inhibiting activity of the molecule (Laurell, 1960; Lancet, 1961; Belyavin, 1963).

Of equal importance is the size of the molecule of the glycoprotein, since a certain minimal dimension of the molecule seems to be necessary for the manifestation of haemagglutination inhibiting activity (GOTTSCHALK, 1960). By means of polymerization, WHITEHEAD (1965), WHITEHEAD et al. (1965), and MORAWIECKI and LISOWSKA (1965) succeeded in demonstrating an increase in the originally low haemagglutination-inhibiting activity of orosomucoid towards some strains of myxoviruses. The inhibiting action against PR 8 influenza virus was increased by a factor of 1,600; with other strains this factor was lower. SPRINGER (1967) pointed out that the inhibition of the haemagglutination by some myxoviruses depended largely on the molecular size of the glycoprotein tested. The haemagglutination-inhibiting and blood-group activities were highest with the largest molecules.

Using electrophoresis on starch, TYRELL (1954) found alpha-inhibitor in the mucoprotein fraction of the alpha-globulins of rabbit sera. The alpha-inhibitor activity of human sera was bound to $alpha_1$- and $alpha_2$-globulins. However, BÜRGI et al. (1961) could not demonstrate any haemagglutination-inhibiting activity in pure $alpha_2$-glycoprotein prepared from human plasma. According to these authors, the residual haemagglutination-inhibiting activity in $alpha_2$ fraction could have been due to traces of the $alpha_1$ acid glycoprotein. Using electrophoresis on starch, HARBOE et al. (1958) also found the maximal haemagglutination-inhibiting activity in the fraction containing both albumins and fast-moving globulins. With the exception of pure albumins and pure gamma-globulins, all remaining fractions still showed an appreciable activity. These authors are of the opinion that, in rabbit serum, various glycoproteins of different electrophoretic mobility may display alpha-inhibitor activity.

The first successful purification of the alpha-inhibitor was achieved by STULBERG et al. (1951) who also showed its mucoprotein nature. HARBOE et al. (1958) argued against their results, stating that, after electrophoresis of the pure mucoprotein, STULBERG et al. made no attempt to elute the particular fraction and to test it for haemagglutination-inhibiting activity.

Several purified glycoproteins isolated from human sera were assayed for their haemagglutination-inhibiting activity by LAURELL (1960). She found haptoglobulin and coeruloplasmin to be the most active inhibitors of influenza A and B viruses. On the other hand orosomucoid inhibited haemagglutination only when heated virus was used. Moreover, high concentrations of $alpha_2$-macroglobulin were necessary to demonstrate its haemagglutination-inhibiting activity. The latter finding is in disagreement with the results of BIDDLE et al. (1967), who found an appreciable part of the inhibiting activity of human sera in the $alpha_2$-macroglobulins. These authors characterised alpha-macroglobulins as a complex of several glycoproteins, several of which possess biological activity. According to BIDDLE et al. (1967), all inhibitors of influenza viruses could be found in this particular complex.

Based on LAURELL'S (1960) and ODIN'S (1955) findings, the presence of sialic acid in a biologically active glycoprotein cannot be considered crucial for the haemagglutination-inhibiting properties of the glycoprotein. Thus, while a solution of haptoglobulin containing 0.003 μg of sialic acid inhibited 4 haemagglutinating units of influenza virus, a solution of $alpha_2$-macroglobulin containing 30 μg sialic acid produced no inhibition. The authors mentioned suggested that either a specific binding of sialic acid in the molecule is necessary to provide its complementarity to the viral neuraminidase, or that the steric configuration of the prosthetic groups in the inhibitor (receptor) is essential for the binding of the influenza virus particles.

In a similar way the affinity of different strains of influenza virus for the inhibitor is determined mainly by the steric configuration of the viral surface (NICULESCU, 1965). The binding between the inhibitor and the virus is mediated through "specific" groups of both reacting surfaces. An electron-microscopic demonstration of the binding of the inhibitor on the virus surface was offered by BAYER (1965).

A change in the surface structure of influenza virus may be achieved by treatment with certain chemicals or by physical processes, e.g. mild heating of the virus (STONE, 1949), treatment with KIO_4 (BORECKÝ et al., 1962a; STYK, 1962a; KRASOVSKAYA, 1964) or ether (NICULESCU, 1965) or gel filtration (ISACHENKO, 1965). A change in the sensitivity of influenza virus population to the inhibitor is frequently achieved either by passaging the virus in another host (STONE, 1951; BRIODY et al., 1955; BRIODY, 1957a; MEDILL-BROWN et al., 1955; STYK, 1957; COHEN et al., 1960a; WERNER et al., 1960; BORECKÝ et al. 1965) or by passaging the inhibitor-sensitive virus in chick embryos in the presence of the given inhibitor (DAVOLI et al., 1959; FAUCONNIER et al., 1959; ORSI et al., 1967; CHOPPIN et al., 1960b).

The mode of action of alpha-inhibitors may be explained as a result of competition for virus between the inhibitor and the cell-receptor (ANDERSON, 1948; GOTTSCHALK, 1959). The haemagglutination-inhibiting reaction occurs when the affinity of the virus to the inhibitor is greater than its affinity to the cell receptor (SMITH et al., 1950). From this it follows that the titre of an inhibitor depends not only on the virus used, but also on the quality of the erythrocytes used in the reaction (STYK, 1955; REIMOLD, 1959; LOBODZIŇSKA et al., 1965). The alpha-inhibitor is devoid of virus-neutralizing activity. Native influenza virus, like RDE, splits off sialic acid from the inhibitor which results in a loss of the inhibitor's ability to bind the virus. As shown by HOLLÓS (1963), however, the coupling of the diazonium salt with the cyclic amino acid of the peptide part of the alpha-inhibitor results in an alteration of its molecule in such a way that the molecule is able to form a firm complex with the virion, which will inhibit the replication of influenza virus. (The modification resulted in the ability to neutralize the infectivity of influenza virus.)

Beta-Inhibitor

Opinions about the chemical nature of beta-inhibitor (sometimes also called the thermolabile serum inhibitor or the virus-neutralizing factor) are even more confusing than those about the alpha-inhibitor. A number of papers (BURNET et al., 1946; GINSBERG et al., 1949; SMITH, 1951; CHU, 1951; SMORODINTSEV et al., 1951; BRANS et al., 1953; KARZON, 1956) reported on certain thermolabile components found in sera which neutralize the infectivity of various viruses. As shown by LUZYANINA et al. (1960), normal animal sera often neutralize not only several myxoviruses, but also enteric viruses, e.g. polio, ECHO and Coxsackie viruses, and certain strains of adenoviruses. It is not clear whether all these serum substances represent the beta-inhibitor or whether there exist also other components in normal animal sera which could act similarly on other viruses. RATHOVÁ et al. (1963) compared the virus-neutralizing activities of bovine serum and of the purified beta-inhibitor prepared from the same serum against several groups of viruses. They found that different inhibitor must be taken into consideration only in the case of mumps virus. No activity of purified beta-inhibitor against this virus could be detected.

Some authors considered the virus-neutralizing activity of sera to be mediated through the properdin system (WAGNER, 1955; KARZON, 1956, 1960; VOLUYSKAYA et al., 1958; GINSBERG et al., 1960; KARZON, 1960). However, recent experimental evidence supports the view that thermolabile inhibitors of normal animal sera are distinct from the properdin system (KONNO, 1958; RATHOVÁ et al., 1959; LUZYANINA et al., 1961; KRIŽANOVÁ et al., 1963; KRIKSZENS, 1967). For instance, no correlation could be demonstrated between the amount of properdin in different sera and their virus-neutralizing activity. The purified beta-inhibitor did not possess any properdin activity despite its high virus-neutralizing activity. Some properties of both properdin and beta-inhibitor are alike, but in contrast to properdin the beta-inhibitor needs Ca ions but does not need components of complement for its activity (KRIŽANOVÁ et al., 1963).

In this respect the results of TOUSSAINT et al. (1959) may be considered the most convincing. These authors found that after absorption of normal sera with *Escherichia coli*, the amount of properdin decreased 2—3 times, while the phage-neutralizing activity remained unchanged. Subsequent absorption of the serum with phage produced no further decrease in the properdin activity, but the phage-neutralizing activity was completely abolished.

MCCREA (1946) was the first to use either ammonium sulphate or sodium sulphate to purify the beta-inhibitor. He found the beta-inhibitor activity in the gamma-globulin fraction of serum. Another successful purification of beta-inhibitor by means of its adsorption on zymozan was carried out by KONNO (1958). His beta-inhibitor also possessed a high specific virus-neutralizing activity, but he still obtained two peaks on centrifugation. KONNO characterized the beta-inhibitor as an euglobulin whose activity depends on Ca ions and which has an electrophoretic mobility between the gamma- and the beta-globulins. Subsequently several authors tried to isolate and identify beta-inhibitor. LUZYANINA et al. (1958) and POLYAK et al. (1959) considered the beta-inhibitor to be a beta-lipoprotein, while HÁNA et al. (1962) and HÁNA (1963 a) considered it a beta-2M globulin. These discrepancies can be partially explained by the fact that the authors used different isolation procedures. So far the purest preparation of beta-inhibitor from bovine serum was obtained by adsorption on to and elution from influenza A1 virus (a beta-inhibitor-sensitive strain) (KRIŽANOVÁ et al., 1963, 1966; KRIKSZENS, 1967). Purified beta-inhibitor is of protein nature and contains 2—5 % sugar which is not essential for its activity. Buoyant density gradient centrifugation reveals a density of 1.35 (1.3—1.4) (Fig. 1). An analysis in the ultracentrifuge showed the beta-inhibitor to be a homogeneous material sedimenting at 4 S (KRIŽANOVÁ et al., 1963, 1966). When gel filtration on Sephadex G-200 was used a pure beta-inhibitor was eluted between the albumin and ethyldehydrogenase

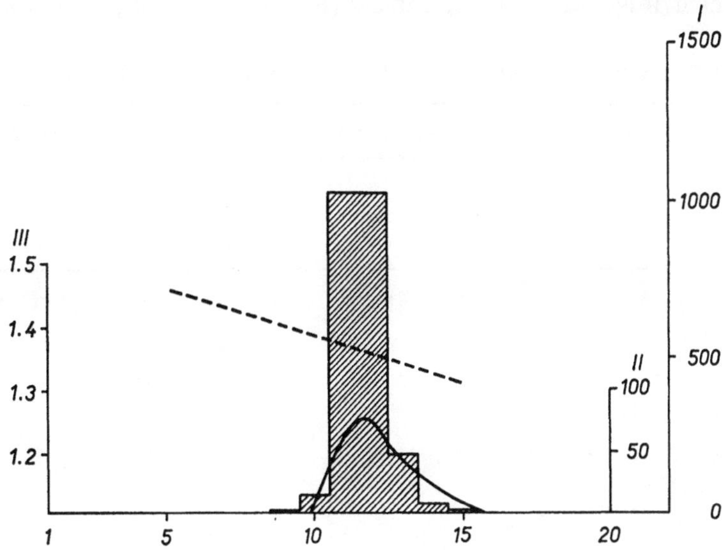

Fig. 1. Equilibrium density gradient centrifugation of bovine serum beta-inhibitor in CsCl solution. Abscissa: fraction number (bottom fraction = 1). Ordinates: *I* HI activity per ml (shaded columns); *II* μg proteins per ml (solid line); *III* density/S_{25} (dashed line). [Reprinted from KRIŽANOVÁ and SOKOL: Acta virol. **10**, 35—42 (1966)]

fractions. But when whole serum was subjected to gel filtration, the beta-inhibitor activity was found in the first peak, e.i. in macroglobulins. Mercaptoethanol did not influence the activity of beta-inhibitor purified by the virus adsorption-elution procedure. However, in the ultracentrifugally separated inhibitor (presumably complexed with macroglobulins), mercaptoethanol enhanced its activity (KRIKSZENS, 1967). Both this activation and the results obtained by gel filtration suggest that beta-inhibitor may be complexed with macroglobulins present in whole serum. This complex may be disrupted by the presence of virus. Based on KRIKSZENS' finding, the inability of the beta-inhibitor-virus complex to fix complement seems to indicate that beta-inhibitor cannot be considered a "classical" antibody, but rather a substance naturally occurring in bovine serum with a capacity to combine with influenza virus. This assumption is supported by the results of BORECKÝ (1967) who found all three types of inhibitor (alpha-, beta-, and gamma-inhibitors) in the serum from a patient with the clinical diagnosis of agammaglobulinaemia. (Such patients, despite the presence of gamma-globulin in their sera, do not respond to antigenic stimulus by formation of detectable antibodies.) As yet no antigen has been found which could raise the level of beta-inhibitor or explain the origin of beta-inhibitor in the organism (BORECKÝ et al., 1963a, b).

An interaction between beta-inhibitor and virus can proceed only in the presence of Ca ions. Complexed with beta-inhibitor, the influenza virus loses its haemagglutinating and infective properties. As a result of the removal of Ca ions by EDTA, the virus-inhibitor complex dissociates into its components and the original activity of both can be recovered after their separation in the ultracentrifuge, as shown in Table 2 (KRIŽANOVÁ et al., 1963, 1966). Beta-

Table 2. *Recovery of beta-inhibitor from its complex with A1-influenza virus*

Sample	Experiment 1			Experiment 2		
	Vol. (ml)	HI units per ml	HA units per ml	Vol. (ml)	HI units per ml	HA units per ml
1. I[a]	13	16,000	0	20	8,000	0
2. V[b]	13	0	32,000	20	0	16,000
3. Supernate after removal of I + V complex	26	0	±512	20	0	256
4. Dissociated I + V complex; 1st supernate	7	±16,000	16	20	4,096	0
5. Dissociated I + V complex; 2nd supernate	7	512	32	20	256	0
6. Virus recovered from I + V complex	6	0	128,000	20	0	16,000

[a] Beta-inhibitor purified by chromatography on silicagel and DEAE-cellulose columns.

[b] Purified A1 influenza virus. [Reprinted from KRIŽANOVÁ and SOKOL, Acta virol. **10**, 35—42 (1966).]

inhibitor does not inhibit the mouse-adapted strain of influenza A1 virus, nor the strain passed in chick embryos in the presence of bovine serum beta-inhibitor (BRIODY et al., 1955a, b; SMORODINTSEV et al., 1960). It may be assumed, therefore, that, as with alpha-inhibitor, a definitive complementary steric configuration of the virus surface is necessary for the reaction of beta-inhibitor with the virus.

Gamma-Inhibitor

The designation "gamma-inhibitor" was proposed by SHIMOJO et al. (1958) for a substance present in normal animal sera which selectively reacts with inhibitor-sensitive strains of influenza A2 virus (MEYER et al., 1957; JENSEN, 1957; CHOPPIN et al., 1959, 1960a, b; JAMES et al., 1959; TAKÁTSY et al., 1959; VOLÁKOVÁ et al., 1959; LUZYANINA et al., 1960b; SHIRATORI et al., 1961;

ZHDANOV et al., 1958). Gamma-inhibitors, like alpha-inhibitors, are of glycoprotein nature. Unlike the alpha-inhibitor, gamma-inhibitor possesses a virus-neutralizing activity.

Electrophoretical studies of gamma-inhibitor (LEVY et al., 1959; HÁNA et al., 1960a; POLYAK et al., 1961; GAMBURG, 1961) confirmed its presence in at least two of the serum protein fractions. Several authors tried to purify gamma-inhibitor and identify its nature (COHEN et al., 1961, 1962, 1965a; SUGIURA et al., 1961a; BELYAVIN et al., 1962; STYK et al., 1962b; KRIŽANOVÁ et al., 1961b, c, 1964; HÁNA et al., 1963b; BORETTI et al., 1964; CASAZZA et al., 1964; BIDDLE et al., 1965). However, their opinions on the nature and size of the gamma-inhibitor molecule are rather contradictory, since molecular sizes from 4S—18S have been reported. Since at least two kinds of gamma-inhibitor may be found in normal sera (KRIŽANOVÁ et al., 1961c; KRIŽANOVÁ and LEŠKO, 1964), it is possible that several authors could have isolated different fractions. This may be an explanation of the discrepancies mentioned above. Such an explanation would be in agreement with the results of HARBOE et al. (1958), who expressed the view that chemically different substances in rabbit serum may have alpha-inhibitor activity. Probably a binding occurs between the biologically active group of the inhibitor with inert "carriers" of different electrophoretic mobility.

Under carefully controlled conditions, BIDDLE et al. (1965) were able to isolate a gamma-inhibitor of high molecular weight (18S), which was extremely sensitive to changes in pH and ionic strength. These authors assume that the gamma-inhibitors of a lower sedimentation coefficient isolated by other workers might result from the breakdown of the 18S molecule. Indeed, in attempts to isolate gamma-inhibitor, different authors used different conditions, which could have resulted in denaturation and thus in changes in gamma-inhibitor molecules.

On the other hand, the possibility exists that gamma-inhibitor occurs in normal animal sera as a complex with other inert materials, as was found for the beta-inhibitor (KRIKSZENS, 1967) and presumably also for the alpha-inhibitor (HARBOE et al., 1958). The results of HÁNA et al. (1961), who found an appreciable increase of gamma-inhibitor activity in guinea pig sera after delipidization, support the hypothesis mentioned above. Such treatment of sera with ether resulted in a more intense staining of alpha$_2$ glycoproteins. After separation of sera by paper electrophoresis, a decrease of lipid staining capacity of this fraction could be also observed. These results may be explained by a release or unmasking of gamma-inhibitor from the lipoprotein complex. DAVOLI et al. (1959) also assume a release of gamma-inhibitor from a complex with other molecules, which occurs after heating the serum and results in activation of the gamma-inhibitor. The lability of the isolated 18S macromolecule possessing gamma-inhibitor activity to both pH and ionic strength (BIDDLE et al., 1965) has no relation with the stability of gamma-inhibitor activity in the whole serum. Using separation on Sephadex G-200, HÁNA et al. (1963b) and STYK et al. (1962b) found gamma-inhibitor activity in sera of newborn unsuckled piglets in the third peak (Fig. 2). These sera usually contain only a small amount of macroglobulins. The same authors, using zonal centrifugation, demonstrated gamma-inhibitor activity in fractions corresponding to substances with a sedimentation coefficient of 3.01—4.03. However, in sera of adult pigs they found gamma-inhibitor activity in macroglobulin fractions (see Fig. 3). The results of the above-mentioned authors might support the hypothesis that gamma-inhibitor is a small molecule bound to macroglobulins in the serum.

The discussion concerning the size of the gamma-inhibitor molecule cannot be concluded until further important questions are clarified, such as the activity of gamma-inhibitor "macromolecule" after dissociation, relation of the size of the molecule to its specific activity, etc.

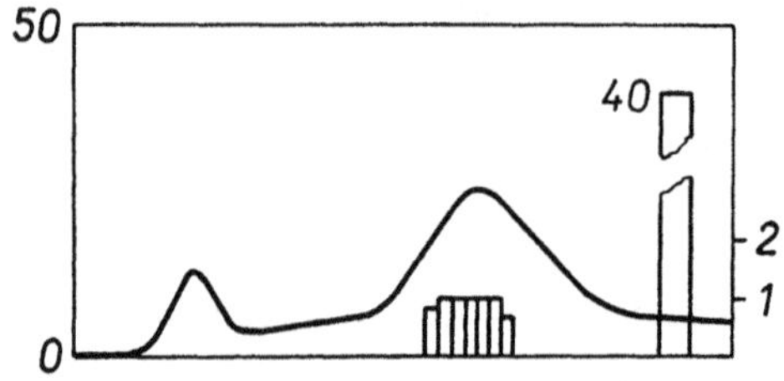

Fig. 2

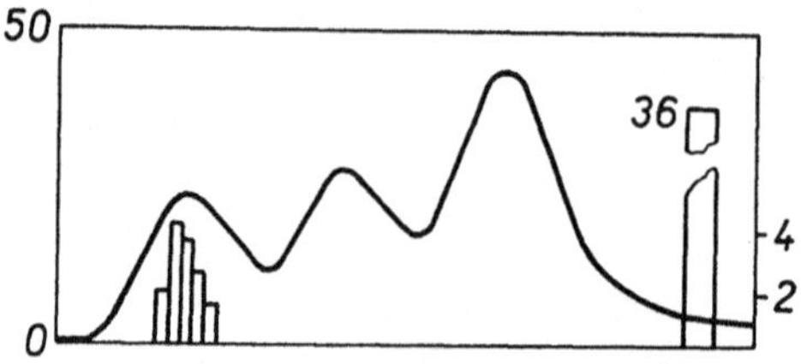

Fig. 3

Fig. 2. Separation on Sephadex G-200 of proteins from serum of newborn unsuckled piglet. Abscissa: fraction number. Left ordinate: absorbance [2537 Å/% (solid line)]. Right ordinate: gamma-inhibitor activity (columns). Column at right: titre determined with whole original serum. [The data in the Fig. 2 are taken from the paper by Hána, Styk, and Schramek: Science Tools **10,** 1—4 (1963)]

Fig. 3. Separation on Sephadex G-200 of proteins from serum of 24-day old piglet. Explanations as in Fig. 2

Unlike alpha-inhibitor, gamma-inhibitor is not completely inactivated by RDE (Choppin et al., 1960a; Cohen et al., 1959; Shimojo et al., 1958; Sugiura et al., 1961b). RDE splits off about 40—50% of sialic acid from the gamma-inhibitor molecule without detectable changes in its haemagglutination-inhibiting activity (Križanová et al., 1961b; Boretti et al., 1964). The rest of the sialic acid, probably responsible for the haemagglutination-inhibiting activity, is neuraminidase-resistant. Such a neuraminidase-resistant molecule of sialic acid was also found in other mucoproteins (Gibbons, 1963; Simkin et al., 1964). Either N-O-diacetyl, or N-acetyl neuraminic acid may be neuraminidase-resistant. These linkages are bound inside the glycoprotein molecule in a way different from alpha-glycosidic bonds (Gibbons, 1963). Under alkaline conditions, the O-acetyl group is split off and the non-sensitive sialic acid becomes sensitive to neuraminidase.

Gibbons found that treatment of bovine submaxillary mucoprotein with 0.05 N Na_2CO_3 at 100° C for 20 minutes resulted in an increase in the amount of neuraminidase-sensitive neuraminic acid from the original 45 % up to 91 %. Similar conclusions were made by Simkin et al. (1964) who studied acid glycoproteins isolated from guinea-pig sera. We assume that experiments of a similar nature might help to elucidate the problem of the active groups of gamma-inhibitor.

SUGIURA et al. (1961b) reported that a decrease in the titre of gamma-inhibitor resulted from treatment with influenza virus enzyme or RDE. However, substantially higher quantities of virus and a longer period of incubation were necessary for gamma-inhibitor inactivation than in the case of ovomucin. The decrease in the sialic acid level of gamma-inhibitor due to the action of virus was quite small in comparison with the resulting decrease in haemagglutination-inhibiting activity. In both haemagglutination-inhibition and virus-neutralization tests, gamma-inhibitor reacts only with inhibitor-sensitive strains of A2 influenza virus (RATHOVÁ et al., 1963), even if the virus is disrupted by ether (WERNER et al., 1960; CHOPPIN, 1964). Besides sensitive and resistant variants of A2 virus, the occurrence of so-called "transitional" strains was described, sensitive to gamma-inhibitor in the haemagglutination-inhibition test but insensitive in the virus-neutralization test (JANDÁSEK, 1964). Gamma-inhibitor may adsorb or flocculate various strains of influenza virus (COHEN et al., 1963b; BIDDLE, 1966). Higher flocculation titres were observed with purified inhibitor. No infectivity of virus-inhibitor complex could be demonstrated. SZÁNTÓ et al. (1962), however, were able to recover fully infective virus by diluting the gamma-inhibitor-virus complex. Similar results were published later by CASAZZA et al. (1964). The dissociation of the complex into its original components probably is not complete, but rather some active groups on the viral surface are released on dilution, allowing the adsorption of virus onto susceptible cells and the initiation of infection. But the inhibitor must not be in excess. Otherwise, a complex of gamma-inhibitor could be formed even with a virus already adsorbed to cell receptors. Under such conditions the virus may show a greater affinity for inhibitor than for cell receptors (CASAZZA et al., 1964). Gamma-inhibitor exerts no effect on intracellular virus.

C Inhibitor

C inhibitor was discovered by STYK (1955) in rat serum. So far no attempts to purify C inhibitor were described. In its properties C inhibitor most closely resembles gamma-inhibitor (HÁNA, 1958; STYK, 1963). Its activity can be increased either by heating or delipidization of the serum (HÁNA et al., 1960b). C inhibitor also is a serum glycoprotein. Its steric configuration seems to be complementary to the surface of the influenza C virus. Alpha-, gamma-, and C inhibitors all appear to be chemically similar substances with different steric configurations and probably containing various amounts and kinds of sialic acid. These differences seem to be responsible for their different affinities to influenza virus strains.

Inhibitors to Other Myxoviruses

In addition to the inhibitors mentioned above, the presence in normal animal sera of non-specific inhibitors against other myxoviruses has been reported.

GINSBERG et al. (1949) found that human, guinea-pig, and mouse sera also inactivate Newcastle disease and mumps viruses. Later WENNER et al. (1952) described another inhibitor of these two viruses. They succeeded in removing this inhibitor from the majority of human sera by chloroform, trypsin or KIO_4 treatment, while an increase of the inhibiting activity of monkey sera against the same viruses was registered after chloroform treatment.

KARZON (1956, 1960) confirmed the presence in mammalian sera of an inhibitor of Newcastle disease virus and concluded that a factor differing from beta-inhibitor could be involved. He named it virus-inactivating substance (VIS) and considered it to be identical with properdin. The presence in human sera of an inhibitor of haemagglutination by mumps virus was described by LUZYANINA et al. (1965). In bovine sera an inhibitor of parainfluenza 3 virus was found (DAWSON, 1963) and a virus-neutralizing factor in human, rabbit, guinea-pig, dog, and horse sera were described. In addition to enteroviruses, they also neutralize the influenza A2 (Japan 305) virus (KLEIN et al., 1962, 1964a, b). In this case gamma-inhibitor could have been involved. Inhibitors of parainfluenza 1, 2 and 3 and influenza B viruses were also found in guinea-pig, rabbit, mouse, dog and rat sera (PICHUSKOV, 1965; TOPCIU et al., 1966, 1967). In the latter case the authors did not exclude the possibility that specific antibody was involved in inhibition.

Methods of Removing Non-Specific Inhibitors from Sera

In the course of time, non-specific inhibitors of different myxoviruses were found in almost all kinds of animal sera tested, although in quantities lower than in typical inhibitor-rich animal sera, such as ferret serum (alpha-inhibitor), bovine serum (beta-inhibitor), or horse serum (gamma-inhibitor). Studies on the distribution of inhibitors in various animal sera have been performed by SAMPAIO (1952), KONNO (1958), SHIMOJO (1958), ZHDANOV (1959), LUZYANINA (1961), COHEN (1963), RATHOVÁ (1965), and ANANTHANARAYAN (1960). We summarize in Table 3 the findings of three authors in which they attempted to quantitate roughly the alpha-, beta-, and gamma-inhibitor capacity of different animal sera.

As seen in Table 3, the inhibitor content of various sera varies greatly. This may be due to actual differences in inhibitor content of the different sera (COHEN, 1963a), and to differences in the test conditions, such as the use of different red blood cells (STYK, 1955; REIMOLD, 1955; LOBODZIŃSKA, 1965). The passage history of the virus strain may alter the viral surface configuration and may affect the results.

There are indications in the literature that the level of inhibitors in sera of various animal species may vary (Table 4). PETROV (1960) reported a seasonal variation in inhibitor level. This assumption is supported by experiments where the level of serum inhibitors *in vivo* was artificially changed. BORECKÝ et al. (1961) reported that, after administration of ethionine to the guinea pig, the level of serum inhibitors of influenza A1 and Newcastle disease

Table 3. *Distribution of three types of influenza virus inhibitors in sera from some animal species*

Serum	Type of inhibitor							
	α			β			γ	
	a	b	c	a	b	c	b	c
Mouse	−	±		+++	++		−	
Hamster	−			±				
Guinea pig	++	+	++	+++	++	++	+++	+++
Rabbit	+	++	+++	++	+	++	+++	++
Ferret	++++	+++		±			+++	
Fowl	+++	+	±	±	+	−	+	+
Swine		+					+++	
Bovine		+	±		+++	++++	+	±
Horse		+	±		+	+	+++	++++
Man		++			+			

[a] Data by SAMPAIO (1952).
[b] Data by SHIMOJO (1958).
[c] Data by RATHOVA (1965).

Table 4. *Variability of inhibitory activity found in individual sera from various animal species against some influenza viruses*

Serum	Virus strain and type				
	B 33 B	FM_1 A 1	Geetha A	Rampf A 2	Shope 15 Swine
Guinea-pig	120—480	60—80	120—480	3,840—15,360	0—60
Monkey	40—480	60—160	30—640	160—5,120	30—80
Rabbit	120—160	30—40	60	1,280—1,920	0—40
Calf	0—2,560	0—30	0—160	(0)—5,120	0
Horse	0—(240)	30—80	0—(60)	2,560—5,120	0—(160)
Sheep	160—240	30—60	120—240	80—120	0
Fowl	0—(30)	0	0	30—60	0

Data collected from several tables presented by ANANTHANARAYAN (1960); Bull. Wld Hlth Org. **22**, 409—420 (1960).

viruses decreased in their sera. On the other hand, by treatment of guinea pigs with certain vitamins (in particular vitamins C and K), RATHOVÁ et al. (1963) observed an increase of the neutralizing activity of normal guinea pig sera together with the rise of complement titres. ROVNOVA et al. (1967) published interesting observations on the fluctuation of the serum inhibitor level during immunization of rabbits with an influenza virus.

The haemagglutination-inhibition test is very often used in diagnostic practice. It need not be emphasized that many difficulties arise in this relatively simple test due to the presence of non-specific inhibitors. These have been found in almost all kinds of normal animal sera. In most cases, without an elaborate analysis, it is impossible to distinguish the non-specific inhibitors from specific antibody, particularly in its early forms.

It has been shown that several kinds of non-specific inhibitor may exist even in the same serum and that the same virus may exist in modifications which react in different degrees with a particular inhibitor in the serum. Thus, it is not surprising that many contradictory data have been published on methods for the removal of non-specific inhibitors. Some basic procedures used for removal of various serum inhibitors must be mentioned.

The use of RDE from *Vibrio cholerae* represents the classical method of removing alpha-inhibitor.

The advantages of the use of either bacterial filtrate or pure enzyme in studies on the reactions of sera with various viral strains were alternatively reported (BURNET et al., 1952; ISAACS et al., 1951; CHU, 1951; MULDER et al., 1952; BRANS et al., 1953; JORDAN, 1954). BORECKÝ (1959) found that a filtrate of a Pneumococcus culture had a similar effect on both red blood cell receptors and alpha-inhibitor as the filtrate of *V. cholerae*. But when crude bacterial filtrates are used, other enzymes (proteinases) could participate in the destruction of serum inhibitors in addition to sialidase. RDE was shown to be ineffective in removing beta-inhibitor. On the other hand, trypsin was shown (SAMPAIO et al., 1953) to be highly effective in disintegration not only of alpha-, but especially of beta-inhibitor in several kinds of normal animal sera. With the method recommended by SAMPAIO et al., no inactivation of specific antibody was observed.

Later on several authors used this method with certain modifications, including trypsin concentration and treatment period (GORBUNOVA et al., 1956; JENSEN et al., 1957; LIPPELT, 1959). However, in 1957 the newly isolated influenza A2 virus was found to react with serum inhibitors which could not be removed from sera by trypsin. RDE was shown to be equally ineffective (ANANTHANARAYAN, 1958). On the other hand, the use of KIO_4, which was employed by BURNET (1948) and MCCREA (1948) to destroy mucoidal virus-inhibiting substances (alpha-inhibitors), was shown to be very effective in removing the so-called gamma-inhibitor from sera. This method has become widely used (GORBUNOVA, 1959; FÜRESZ, 1959; FISET, 1964; TAKÁTSY et al., 1959; LUZYANINA, 1960) and was also recommended by the W.H.O. Expert Committee in the W.H.O. Technical Report Series No. 170 (1959).

Besides the use of RDE, trypsin and KIO_4, other procedures have been proposed for removing non-specific inhibitors from sera. Heating of sera at different temperatures, which was frequently used earlier, was often shown to be insufficient. The level of gamma-inhibitor even rose after heating of sera.

The method of bubbling through CO_2, introduced and used mainly by Soviet authors (FRIDMAN, 1955; GORBUNOVA et al., 1956; ZAKSTELSKAYA et al., 1958; ZHDANOV et al., 1958) is now rarely employed, because it hardly controls the reaction quantitatively and often results in a partial inactivation of specific antibody. GOREV (1958), using a filtrate of *Pseudomonas fluorescens*, was successful in removing thermostabile inhibitors. When a demonstration of early antibodies is not required, serum may be freed of all kinds of known inhibitors by treatment with rivanol, which in fact separates serum gammaglobulins (GEFT, 1963).

Comparative studies by different authors who used a variety of methods to remove inhibitors showed that both quantitative and qualitative differences exist among the inhibitors in sera of different animal species (SHIMOJO, 1958; COHEN et al., 1959; GORBUNOVA et al., 1958, 1959; LIPPELT, 1959; LUZYANINA, 1960; GAMBURG, 1961; RATHOVÁ et al., 1961). On the basis of these results there appears to be no method effective in removing all kinds of inhibitors from every serum. The best results seem to be achieved by a method combining effects of trypsin, heating and KIO_4 (JENSEN, 1958). In this way, inhibitors of a majority of influenza viruses were almost completely removed from several kinds of serum (ANANTHARAYAN, 1960).

Because of its simplicity and efficiency, we shall describe in detail this combined method, which is based on FISET'S data (1964). To one volume of serum is added a half volume of trypsin (Difco, 1:250, 8.0 mg/ml) in phosphate-buffered saline, pH 8.2. At this pH, trypsin is rather unstable but retains its activity for a long time when stored in a refrigerator. The mixture is incubated at 56° C for 30 minutes. As the temperature rises, trypsin is rapidly inactivated. Under these conditions the inhibitors are destroyed but the antibodies are not affected. The mixture is then cooled for a few minutes and three volumes of M/90 KIO_4 are added. The mixture is kept at room temperature for 15 minutes and then neutralized with 3 volumes of either a glycerol-saline or glucose-saline solution. To reach final dilution 1:10, 2.5 volumes of saline are added.

However, the destruction of non-specific inhibitors in sera before their examination for antibody activity is disadvantageous, because it has not been shown that these methods do not affect specific antibodies. HORVÁTH (1961) reported a decrease in specific antibodies after 7 minutes' treatment with trypsin (0.1 % at 37° C with subsequent inactivation of trypsin at 57° C for one hour). Partial destruction of antibodies resulting from KIO_4 treatment was observed by HARBOE (1959). Similarly NAGASE (1961) described a partial destruction of antibody after KIO_4 treatment of human sera. YAKHNO (1961) reported a decrease in or even destruction of primary haemagglutination-inhibiting antibodies, but not of antihaemagglutinins in sera obtained after hyperimmunization of rats and guinea pigs with influenza A2 virus, when the sera were treated with both CO_2 and KIO_4. However, virus-neutralizing antibody formed even though the first administration of virus was not affected by this method.

The only way to avoid the risk of either incomplete removal of serum inhibitors, or of destruction of specific antibodies in serological investigations is to use viral strains insensitive to serum inhibitors. These may occur "naturally" or after change by some artificial treatment. The means for changing the sensitivity of virus strains to inhibitors were mentioned above (passaging in the presence of inhibitors, etc.). Based on the application of ALEXANDROVA'S (1962) technique for such viruses, a similar method providing greater reliability of serological results was recommended by NICULESCU (1965). But this method is rather complicated.

A Possible Role of Serum Inhibitors in the Natural Resistance of the Organism

Different approaches have been undertaken to explain the possible role of serum inhibitors in the natural resistance of the organism.

In the first approach an effort has been made to find a relationship between the susceptibility of experimental animals to a particular virus on the one hand, and the neutralizing activity of their sera on the other (BANG et al., 1950). SAWICKI et al. (1963) concluded that guinea pigs and white rats cannot be infected with several strains of influenza virus when natural serum inhibitors against these strains are present in their sera.

This question was dealt with in more detail by SMORODINTSEV et al. (1951) and SMORODINTSEV (1960). They reported that native sera from uninfected white mice possess high virus-neutralizing activity against non-pathogenic strains, which results in an inapparent infection in these animals. However, the mouse sera had no effect on the virulent strains of influenza virus. In the course of the adaptation to mouse lungs, the non-pathogenic strains of influenza virus increase in virulence with a parallel decrease in sensitivity to thermolabile serum inhibitors. This is in accordance with the experiments of BRIODY et al. (1955a, b) and others, where a change of susceptibility of the virus to inhibitor during adaptation to mice was observed (see above, the section dealing with alpha-inhibitor). SMORODINTSEV postulated a protective role of thermolabile inhibitors in viral infection, and a direct role in overcoming the susceptibility of virus to the inhibitor during changes from apathogenic to pathogenic variants.

The second approach was expressed in an effort to influence the level of serum inhibitors *in vivo*. BORECKÝ et al. (1961, 1962b, c) treated experimental animals with ethionine and achieved a decrease in the inhibitory titres against Newcastle disease and influenza A1 virus, but not against the A2/Singapore/57 strain sensitive to the inhibitor. These authors obtained similar results with hamsters. During ethionine treatment of hamsters, the virus-neutralizing activity of their sera against influenza A1 virus was appreciably decreased. Simultaneously with the decrease of the virus-neutralizing activity of hamster sera, both faster multiplication and higher titres of influenza A1 virus adapted to mice (but not of hamster-adapted virus) were observed in the lungs of hamsters fed with ethionine than in those of the controls (Figs. 4, 5).

In this way BORECKÝ et al. succeeded in finding a relationship *in vivo* between the level of serum inhibitors and influenza virus multiplication. These results support the opinions which consider the presence of non-specific inhibitors in serum to be a factor in natural resistance.

The third and most frequently used approach towards an elucidation of the possible role of inhibitors in virus infection consisted in searching for a direct effect of inhibitors on either virus or viral infection. Experimental animals were given inhibitor (either in the form of whole serum or purified inhibitor substance) at different intervals before and after inoculation with

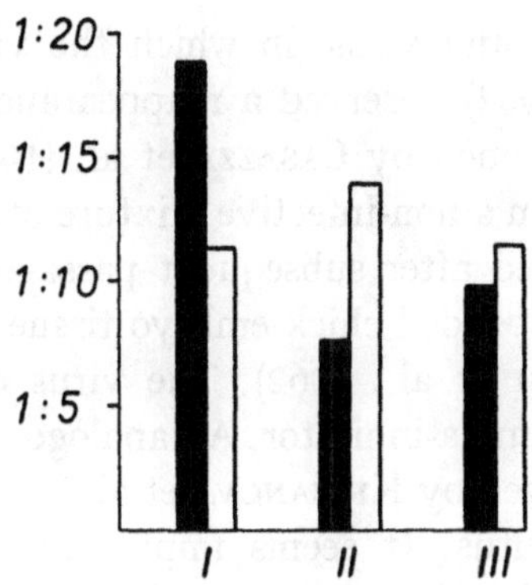

Fig. 4

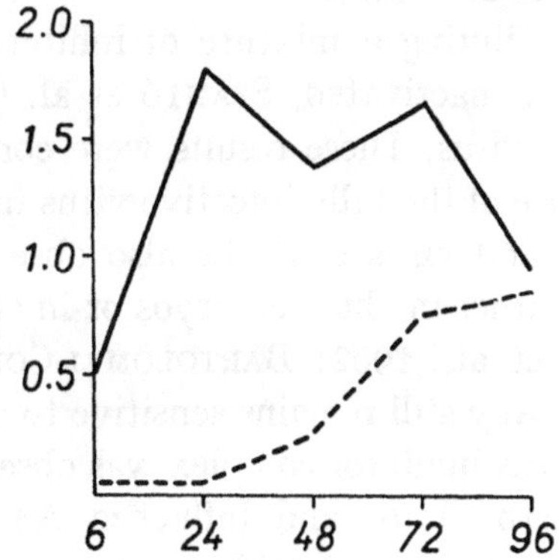

Fig. 5

Fig. 4. Decrease of virus neutralizing activity of normal hamster sera against A1-3711 influenza virus after administration of ethionine. Mean values from 5 examinations of pooled sera from 4—6 hamsters. Black columns: experimental hamsters. White columns: control hamsters. Serum samples collected before ethionine administration (*I*) and after 5 (*II*) and 10 (*III*) injections of a 2.5% ethionine solution. Ordinate: highest serum dilution neutralizing 1,000 $TCID_{50}$ of virus. [Reprinted from BORECKÝ, RATHOVÁ, and KOČIŠKOVÁ: Acta virol. **6,** 97—104 (1962)]

Fig. 5. Infectivity titres of A1-3711 influenza virus in lungs from ethionine-treated (——) and control (---------) hamsters. Mean values from 4 determinations. Abscissa: hour after inoculation. Ordinate: log. $TCID_{50}$/0.1 ml. [Reprinted from BORECKÝ, RATHOVÁ, and KOČIŠKOVÁ: Acta virol. **6,** 97—104 (1962)]

virus. Probably because of its stability, gamma-inhibitor was most often used. It may be concluded that inhibitor inhibits the reproduction in mice lungs of the inhibitor-sensitive strain of influenza A2 virus (DAVOLI et al., 1959; COHEN, 1960b; RYKOWSKA et al., 1963; CASAZZA, 1964; LINK et al., 1964, 1965). The inhibition of virus multiplication could be observed only when a sufficient concentration of inhibitor was given to mice either before, simultaneously with, or shortly after inoculation with virus, i.e. at a time when the virus had not yet penetrated into susceptible cells. Usually both gamma-inhibitor and virus were administered to mice by the same route (i.e. intranasally).

CASAZZA et al. (1964) found that one intranasal dose of gamma-inhibitor 4—6 hours after infection or later, did not influence the course of infection, while high doses of inhibitor given intraperitoneally even at longer intervals after infection resulted in an appreciable virus-neutralizing effect. There is so far no satisfactory explanation for this finding.

The sensitivity of chorioallantoic membranes to viral infection was not changed, provided that membranes were first incubated in the presence of gamma-inhibitor which was then removed just before the application of virus. It seems clear from this experiment that the inhibitor acts on extracellular virus but not on the cell (CASAZZA, 1964; BARTOLOMEI-CORSI, 1962). The effect of the inhibitor on the virus, i.e. the formation of a complex, is manifested by either formation of precipitates (SZÁNTÓ et al., 1962) or flocculation (BIDDLE et al., 1966).

After diluting a mixture of inhibitor and virus, in which the virus was completely inactivated, SZÁNTÓ et al. (1962) observed a reappearance of the infectious virus. These results were confirmed by CASAZZA et al. (1964). Reappearance of the fully infective virus from a non-infective mixture of gamma-inhibitor and virus could be also observed after subsequent passages of the mixture either in chick embryos or in suspended chick-embryo tissue cultures (SZÁNTÓ et al., 1962; BARTOLOMEI-CORSI et al., 1962). The virus obtained in such a way still remains sensitive to gamma-inhibitor. An analogous formation of virus-inhibitor complex was observed by KRIŽANOVÁ et al. (1963, 1966) with beta-inhibitor and influenza A1 virus. It seems important that the presence of Ca ions should be necessary for the formation of the complex.

After removal of Ca ions, e.g. with EDTA, the complex dissociated into its components. Both components reverted to normal conditions in the presence of Ca ions, i.e. they showed the original activity (Table 2). On the basis of these results it could be concluded that the mechanism of the action of serum inhibitors on sensitive variants of viruses resembles, at least in the first reaction step, the interaction of specific antibody with a virus. These inhibitors may bind to complementary places on the viral surface, thus blocking the adsorption of the virus particle onto the receptors of susceptible cells. The internal structure of the virus particle necessary for initiating virus synthesis in a cell remains intact. Unveiling active sites of viral surface by dilution of the virus-inhibitor complex, or removing Ca ions, restores the capacity for adsorbing virus onto susceptible cells and for initiating the infective process.

In COHEN's (1965b) opinion, gamma-inhibitor cannot play a significant role in the natural resistance of man since it is present not at all or only at low titres in human sera.

The protection from viral infection through administration of gamma-inhibitor to experimental animals led many authors to optimistic expectations of the utilization of gamma-inhibitor for prophylactic or therapeutic purposes (BARTOLOMEI-CORSI, 1962; CASAZZA, 1964; JANDÁSEK, 1967). The results of KOSYAKOV et al. (1962a, b, 1963) are instructive in this respect. They compared the therapeutic effect of both specific antibody and non-specific gamma-inhibitor in chick embryos and mice infected with influenza viruses. Their experiments revealed that both the immune antiviral and normal guinea-pig sera (containing gamma-inhibitor) showed an appreciable therapeutic effect when given to animals in the early stages of infection (i.e. 3—24 hours after infection). The presence of the virus in allantoic fluids of infected chick embryos could not be demonstrated, while the virus present in either chorioallantoic membrane cells, or in mice lungs, "remained intact", i.e. it could be recovered by subsequent passages in chick embryos. Such a late recovery of the virus was considered by the authors mentioned as an "incomplete therapeutic effect".

We think that release of a virus from an inactive complex with inhibitors could explain the phenomenon in a sense mentioned previously. Since there is as yet no evidence for the formation of serum inhibitors as a result of an antigenic stimulus, but the mechanism of their interaction with virus is similar to that of specific antibody, it still appears justified to consider serum inhibitors as one of the factors of the natural resistance. However, the possibility of putting them to practical use remains open. The unpredictable change of sensitive influenza viruses into insensitive strains, as well as the possibility of the liberation of fully infective virus from inactive complexes with inhibitors gives little hope for the utilization of inhibitors for therapeutic purposes in the near future.

References

ALEXANDROVA, G. I.: An analysis of the structure of the populations of inhibitor-sensitive and inhibitor-resistant A2 influenza virus strains. Acta virol. **6**, 487—497 (1962).

ANANTHANARYAN, R., and C. J. K. PANIKER: Non-specific inhibitors of influenza viruses in normal sera. Bull. Wld Hlth Org. **22**, 409—420 (1960).

ANDERSON, S. G.: Mucins and mucoids in relation to influenza virus action. Aust. J. exp. Biol. med. Sci. **26**, 347—411 (1948).

BANG, F. B., M. FOARD, and D. T. KARZON: The determination and significance of substances neutralizing Newcastle Virus in human serum. Johns Hopk. Hosp. Bull. **87**, 130—143 (1950).

BARTOLOMEI-CORSI, O., A. M. CASAZZA, A. DI MARCO, M. GHIONE, and A. ZANELLA: Biological properties of a horse serum inhibitor against influenza A2 viruses. In VIIIth Int. Congr. Microbiol., Abstracts, Montreal, Quebec, Canada, No. D. 30. 5p. 96 (1962).

BAYER, M. E.: The reaction of receptor glycoprotein with influenza virus and neuraminidase. An electron microscopic study. Trans. N.Y. Acad. Sci. **26**, Suppl. 1103—1111 (1965).

BELYAVIN, G.: The mechanism of influenza virus haemagglutination as a model of specific receptor adsorption. J. gen. Microbiol. **32**, 15—18 (1963).

—, and A. COHEN: Chromatography of influenza virus haemagglutination inhibitors. Virology **18**, 144—145 (1962).

BIDDLE, F., D. S. PEPPER, and G. BELYAVIN: Properties of horse serum gamma inhibitor. Nature (Lond.) **207**, 381—383 (1965).

—, and K. F. SHORTRIDGE: Immunological cross-reactions of influenza virus inhibitors. Brit. J. exp. Path. **48**, 285—293 (1967).

—, and J. P. STEVENSON: Flocculation of influenza viruses by horse serum inhibitor. Nature (Lond.) **209**, 1223—1225 (1966).

BLAŠKOVIČ, D., N. A. MAXIMOVIČ, B. STYK, and P. ALBRECHT: Course of adaptation of the inhibitor-resistant influenza virus A2 to ferrets. Čs. Epidem. **10**, 158—165 (1961) [in Slovak].

BORECKÝ, L.: A factor destroying virus receptors in pneumococcal cultures. I. Isolation and properties of the factor. Acta virol. **2**, 73—83 (1958).

— Mechanisms of nonspecific resistance of the organism. Bratislava, Vydavatel'stvo Slovenskej akadémie vied 1967 [in Slovak].

— S. HALPEREN, and W. W. ACKERMANN: Comparative study of the interaction of polyoma and other hemagglutinating viruses with an isolated bovine mucoprotein. J. Immunol. **89**, 752—758 (1962a).

— D. KOČIŠKOVÁ, and L. HÁNA: An attempt to affect the level of serum inhibitors of myxoviruses in vivo. I. Effect of ethionine administration on the level of inhibitors in guinea-pig sera. Acta virol. **5**, 236—244 (1961).

— V. LACKOVIČ, and V. RATHOVÁ: Surface properties of the egg and mouse lines of A1 influenza virus. J. Immunol. **95**, 387 (1965).

— — —, and E. MRENA: The syndrome of adaptation of influenza viruses to laboratory animals. J. Hyg. Epidem. (Praha) **10**, 277—287 (1966).

—, and V. RATHOVÁ: Investigations of possible cross reactions between preparations of Salmonellae containing Vi antigen and inhibitors or antibodies against myxoviruses. Acta virol. **7**, 339—349 (1963b).

— —, and D. KOČIŠKOVÁ: An attempt to affect the level of serum inhibitors of myxoviruses in vivo. II. Effect of ethionine administration on virus multiplication. Acta virol. **6**, 97—104 (1962c).

— — —, and L. HÁNA: Effect of d-l ethionine on some factors of nonspecific resistance to myxoviruses. J. Hyg. Epidem. (Praha) **6**, 65—70 (1962b).

— —, and O. KRIŽANOVÁ: Nonspecific resistance of the organism and myxovirus inhibitors. (Comparison of some antibacterial and antiviral mechanisms.) Folia microbiol. (Praha) **8**, 137—146 (1963a).

BORETTI, G., A. DI MARCO, and J. PIERA: Preparation and characterization of an influenza virus inhibitor from horse serum. G. Microbiol. **12**, 55—70 (1964).

BOWARNICK, M., and P. M. DE BURGH: Virus haemagglutination. Science **105**, 550 (1947).

BRANS, L. M., E. HERTZBERGER, and J. L. BINKHORST: Studies on the elimination of nonspecific inhibitor in sera against influenza viruses with the aid of filtrates of *Vibrio Cholerae*. Antonie v. Leeuwenhoek **19**, 309 (1953).

BRIODY, B. A.: Antigenic mutants of influenza virus. Fed. Proc. **16**, 407—408 (1957).

—, and W. A. CASSEL: Adaptation of influenza virus to mice. I. Changes in the growth curve of an A prime strain of influenza virus. J. Immunol. **74**, 37—40 (1955b).

— —, and M. A. MEDILL: Adaptation of influenza virus to mice. III. Development of resistance to beta-inhibitor. J. Immunol. **74**, 41—45 (1955a).

BÜRGI, W., and K. SCHMID: Preparation and properties of Zn-alfa$_2$-glycoprotein of normal human plasma. J. biol. Chem. **236**, 1066—1077 (1961).

BURNET, F. M., and J. D. STONE: The receptor-destroying enzyme of *V. Cholerae*. Aust. J. exp. Biol. med. Sci. **25**, 227—235 (1947).

BURNET, J. M.: The effect of periodate on the virus inhibitory qualities of mucoids in solution. Aust. J. exp. Biol. med. Sci. **27**, 218 (1949).

—, and J. F. MCCREA: Inhibitory and inactivating action of normal ferret sera against an influenza virus strain. Aust. J. exp. Biol. med. Sci. **24**, 277—282 (1946).

CASAZZA, A. M., A. DI MARCO, M. GHIONE, and A. ZANELLA: Biological properties of a horse serum inhibitor against influenza A2 viruses. G. Microbiol. **12**, 1—14 (1964).

CHOPPIN, P. W., and W. STOECKENIUS: Interactions of ether-disrupted influenza A2 virus with erythrocytes, inhibitors, and antibodies. Virology **22**, 482—492 (1964).

—, and I. TAMM: Two kinds of particles with contrasting properties in influenza A virus strains from the 1957 pandemic. Virology **8**, 539—542 (1959).

CHOPPIN, P. W., and W. STOECKENIUS: Studies of two kinds of virus particles which comprise influenza A2 virus strains. II. Reactivity with virus inhibitors in normal sera. J. exp. Med. **112**, 921—944 (1960a).

— — Studies of two kinds of virus particles which comprise influenza A2 virus strains. I. Characterization of stable homogeneous substrains in reactions with specific antibody, mucorpotein inhibitors and erythrocytes. J. exp. Med. **112**, 895—920 (1960b).

CHU, C. M.: The action of normal mouse serum on influenza virus. J. gen. Microbiol. **5**, 739—757 (1951).

COHEN, A.: Protection of mice against Asian influenza virus infection by a normal horse serum inhibitor. Lancet **1960**b **II**, 791—794.

—, and G. BELYAVIN: Hemagglutination inhibition of Asian influenza viruses: A new pattern of response. Virology **7**, 59—74 (1959).

— — The influenza virus hemagglutination inhibitors of normal rabbit serum. I. Separation of the inhibitory components. Virology **13**, 58—67 (1961).

— —, and E. NEWLAND: Analysis of the horse serum inhibitor of A2 influenza virus haemagglutination and infectivity. VIIIth Int. Congr. Microbiol., Montreal, Quebec, Canada, No. D. 30. 4, p. 95 (1962).

—, and F. BIDDLE: The effect of passage in different hosts on the inhibitor sensitivity of an Asian influenza virus strain. Virology **11**, 458—473 (1960a).

— — Haemagglutination inhibition. Interaction of influenza viruses with horse serum inhibitor. Nature (Lond.) **198**, 508—509 (1963b).

— —, and S. E. NEWLAND: Analysis of horse serum inhibitors of A2 influenza virus haemagglutination. Brit. J. exp. Path. **46**, 497—513 (1965a).

—, and J. DORMAN: Non-specific inhibitors of A2 influenza virus in human sera. Acta virol. **9**, 519—525 (1965b).

— S. E. NEWLAND, and F. BIDDLE: Inhibition of influenza virus haemagglutination: A difference of behavior in sera from a single species. Virology **20**, 518—529 (1963a).

DAVOLI, R., and O. BARTOLOMEI-CORSI: Nonspecific inhibition of the haemagglutination by influenza viruses. VI. Haemagglutination-inhibition and neutralization of infectivity of influenza viruses of the type A2 by normal guinea pig serum activated by heat. Sperimentale **109**, 32—42 (1959).

DAWSON, P. S.: The nature of substances present in normal bovine sera inhibiting the activity of parainfluenza 3 virus. J. comp. Path. **73**, 428—436 (1963).

Expert committee on respiratory virus diseases. Wld Hlth Org. techn. Rep. Ser., No 170 (1959).

FAUCONNIER, B., R. M. LARTITÉGUI et D. BARUA: Adaptation artificielle du virus grippal aux inhibiteurs du serum normal. I. Obtention, à partir d'une souche sensible d'une variante insensible par culture du virus en présence de son inhibiteur. Ann. Inst. Pasteur **97**, 473—478 (1959).

FISET, P.: Serological techniques. In: Techniques in experimental virology. London and New York: R. J. C. Harris 1964.

FRANCIS, T., JR.: Dissociation of hemagglutinating and antibody measuring capacities of influenza virus. J. exp. Med. **85**, 1—7 (1947).

FRIDMAN, E. A.: Non-specific neutralization of influenza virus and methods for its removal. Thesis Leningrad 1955 [in Russian].

FÜRESZ, J.: Studies on inhibitors of influenza virus haemagglutination in normal rabbit and ferret sera. Canad. J. Microbiol. **5**, 505—511 (1959).

GAMBURG, V. P.: Haemagglutination inhibitor and neutralizing factor of A2 influenza virus in sera. Acta virol. **5**, 317—324 (1961).

GEFT, R. A., and R. YA. POLYAK: Removal of thermostable inhibitors against A2 influenza virus from immune horse sera by rivanol. Acta virol. **7**, 91 (1963).

GIBBONS, R. A.: The sensitivity of the neuraminosidic linkage in mucosubstances towards acid and towards neuraminidase. Biochem. J. **89**, 380—391 (1963).
GINSBERG, H., and R. J. WEDGWOOD: Properdin and the inactivation of viruses. Ann. N.Y. Acad. Sci. **83**, 528—532 (1960).
GINSBERG, H. S., and F. L. HORSFALL: A labile component of normal serum which combines with various viruses. Neutralization of infectivity and inhibition of hemagglutination by the component. J. exp. Med. **90**, 475—495 (1949).
GORBUNOVA, A. S., V. M. STAKHANOVA, A. N. LOZHKINA, and V. D. OLLI: Comparative efficacy of methods for treatment of sera with carbon dioxide, *V. Cholerae* filtrate and potassium periodate for eliminating unspecific inhibitors of hemagglutination to influenza A2 virus. Vop. Virus. **4**, 750—753 (1959) [in Russian].
GOREV, N. E.: The use of broth culture filtrates of *Pseudomonas fluorescens* for the destruction of non-specific thermostabile (at 56° C) influenza virus inhibitor in human sera. Acta virol. **2**, 171—178 (1958).
GOTTSCHALK, A.: Chemistry of virus receptors. In: The viruses (F. M. BURNET and W. M. STANLEY). New York and London 1959.
— The chemistry and biology of sialic acids and related substances. Cambridge 1960.
HÁNA, L.: Some properties of haemagglutination inhibitor against influenza C virus. Proc. of the IVth Int. Congr. Biochem. Vienna 1958, **7**, Sympos. 7, 160—162 (1958).
— Non-specific inhibitors of myxoviruses. Zprávy z epidemiol. a mikrobiol. **5**, 40—43 (1963a) [in Slovak].
—, and B. STYK: Influence of delipidisation on the haemagglutination inhibiting activity of rat serum against influenza typ C viruses. Acta virol. **4**, 392—393 (1960b).
— — Lipoids as inhibitors of heamagglutination-inhibiting activity against "avid" of A2 influenza. Acta virol. **5**, 190—192 (1961).
— — An attempt to differentiate beta-inhibitor from cofactor occurring in bovine serum. Acta virol. **6**, 77—83 (1962).
— —, and D. KOČIŠKOVÁ: Estimation of the distribution and the character of inhibitors against "avid" A2 influenza virus strains in protein fractions of guinea pig and horse sera obtained by paper electrophoresis. Acta virol. **4**, 356—364 (1960a).
— —, and Š. SCHRAMEK: Studies on the size of antibodies, cofactor and nonspecific inhibitors of influenza virus using separation on Sephadex G-200. Science Tools **10**, 1—4 (1963b).
HARBOE, A., and R. REENAAS: The course of antibody development to influenza A2 strains. Acta path. microbiol. scand. **46**, 266—272 (1959).
— —, and M. OPPEDAL: Studies on Francis inhibitor of influenza virus hemagglutination in rabbit serum fractions obtained by electrophoresis on starch-grain. Acta path. microbiol. scand. **44**, 92—105 (1958).
HOLLÓS, I.: The effect of modified Francis-inhibitors on reproduction of influenza virus. Acta microbiol. Acad. Sci. hung. **10**, 82—83 (1963).
HORVÁTH, S.: Comparison of the experimental results obtained in the course of inactivation of non-specific inhibitors in serum by means of the haemagglutination and virus neutralization tests. Arch. ges. Virusforsch. **11**, 295—303 (1961).
ISAACS, A., and A. BOZZO: The use of *V. cholerae* filtrates in the destruction of non-specific inhibitor in ferret sera. Brit. J. exp. Path. **32**, 325—335 (1951).
ISACHENKO, V. A., G. A. SMIRNOVA, and A. S. GORBUNOVA: Isolation of an inhibitor-sensitive variant from an inhibitor-resistant influenza A2 culture by gel-filtration with Sephadex. Vop. Virus. **10**, 97—98 (1965) [in Russian].
JAMES, S. M., and P. FISET: Serum inhibitors of Asian strains of influenza virus. Nature (Lond.) **184**, Suppl. 21, 1656—1657 (1959).

JANDÁSEK, L.: Study of the interaction between the "transition strains" of influenza virus A2 and the inhibitor contained in the sera of normal guinea pigs. J. Hyg. Epidem. (Praha) **8**, 44—48 (1964).

— Inhibitoren des Influenzavirus. Arch. exp. Vet. Med. **21**, 313—318 (1967).

JENSEN, K. E.: "Memorandum". Communicable Disease Center, Montgomery, Alabama, 1958.

—, and A. MONTGOMERY: New set of type A influenza viruses. J. Amer. med. Ass. **164**, 2025—2029 (1957).

JORDAN, W. S., and R. O. OZEASOHN: The use of RDE to improve the sensitivity of hemagglutination-inhibition test for the serologic diagnosis of influenza. J. Immunol. **72**, 229—235 (1954).

KARZON, D. T.: Non-specific viral inactivating substance (VIS) in human and mammalian sera. J. Immunol. **76**, 454—463 (1956).

—, and R. H. BUSSELL: Studies of the nonspecific inactivation of Newcastle disease virus by mammalian sera. Ann. N.Y. Acad. Sci. **83**, 539—550 (1960).

KATHAN, R. H., C. A. JOHNSON, and R. J. WINZLER: On the glycoprotein inhibitor of viral hemagglutination from human erythrocytes. Acta Soc. Med. upsalien. **64**, 290 (1959).

—, and R. J. WINZLER: Structure studies on the myxovirus hemagglutination inhibitor of human erythrocytes. J. biol. Chem. **238**, 21—25 (1963).

— —, and C. A. JOHNSON: Preparation of an inhibitor of viral hemagglutination from human erythrocytes. J. exp. Med. **113**, 37—45 (1961).

KLEIN, M., and A. DEFOREST: A serum inhibitor in animal and human sera of bovine enteroviruses. II. In vivo activity and role in "natural" resistance to infection. J. Immunol. **92**, 822—826 (1964b).

— G. EVANS, A. DEFOREST, and B. KLEGER: A serum inhibitor in animal and human sera of bovine enteroviruses. I. Properties and in vitro activity. J. Immunol. **95**, 816—821 (1964a).

— —, and Y. ZELLAT: Properties of non-antibody inhibitor in the sera of man and other animals. Fed. Proc. **21**, 461 (1962).

KONNO, J.: Beta-inhibitor, its biological and physico-chemical properties with particular emphasis on the differences from alfa-inhibitor, immune serum and properdin. Tohoku J. exp. Med. **67**, 391—405 (1958).

KOSYAKOV, P. N., M. S. BERDINSKIKH, and Z. I. ROVNOVA: Ability of viruses to overcome the action of inhibitors and antibodies. Vop. Virus. **7**, 28—35 (1962b) [in Russian].

— N. M. GRUZDEVA, and M. S. BERDINSKIKH: The problem of the therapeutic effect of specific antibodies and inhibitors in influenza infection. Vop. Virus. **8**, 301—307 (1963) [in Russian].

—, and Z. I. ROVNOVA: Virus-cell system exposed to specific antibody and inhibitors. VIIIth Int. Congr. for Microbiology, Abstracts, Montreal, Quebec, Canada, No. E 32 B. 12. p. 109 (1962a).

KRASOVSKAYA, I. A.: Changes in some properties of influenza A2 virus after treatment with sodium periodate. Vop. Virus. **9**, 434—437 (1964) [in Russian].

KRIKSZENS, A. E.: Some chemical and physical characteristics of purified beta-inhibitor from bovine serum. Proc. Soc. exp. Biol. (N.Y.) **126**, 176—181 (1967).

KRIŽANOVÁ, O., and J. LEŠKO: Separation of gamma-inhibitor on carboxymethyl cellulose. Acta virol. **7**, 179—182 (1964).

— V. RATHOVÁ, D. KOČIŠKOVÁ, and J. SZÁNTÓ: The purification and some properties of the beta-inhibitor from bovine serum. Bratisl. lek. Listy **43**, 22—29 (1963) [in Slovak].

—, and F. SOKOL: Purification of bovine serum beta inhibitor by chromatographic methods and its reaction with A1 influenza virus. Acta virol. **10**, 35—42 (1966).

Križanová-Laučíková, O., J. Szántó, and P. Albrecht: Localization of virus hemagglutination inhibitor in the chorioallantoic membrane. Virology **15**, 501—503 (1961a).

— —, and D. Kočišková: Purification and some properties of the thermostable inhibitor against avid A2 influenza virus from horse serum. Acta virol. **5**, 4—11 (1961b).

— — — G. Ruttkay-Nedecký, and F. Sokol: Differences in the properties of two inhibitors avid A2 influenza virus strains from horse serum. Acta virol. **5**, 12—18 (1961c).

Laurell, A. B.: Inhibitor capacity of some purified human serum proteins on haemagglutination by influenza virus. Acta path. microbiol. scand. **49**, 213—222 (1960).

Levy, A. H., P. S. Norman, and R. R. Wagner: Electrophoresis of serum inhibitors of influenza viruses. Fed. Proc. **18**, 580 (1959).

Link, F., D. Blaškovič, and J. Raus: On the therapeutic effect of gamma inhibitor in mice infected with an inhibitor-sensitive unadapted A2 influenza virus. Acta virol. **9**, 553 (1965).

— J. Szántó, and O. Križanová: A quantitative assay of the in vivo protective effect of gamma inhibitor against inhibitor sensitive A2 influenza virus strain. Acta virol. **8**, 71—75 (1964).

Lippelt, H., u. W. Wirth: Unspezifische Hemmfaktoren im Meerschwein-Normalserum gegenüber Influenza-Viren. Arch. ges. Virusforsch. **9**, 497—509 (1959).

Lobodzińska, M., and O. Križanová: The relationship between hemagglutination and adsorption of influenza viruses on red blood cells of different animal species and their total sialic acid content. Arch. Immunol. Ter. dośw. **13**, 631—638 (1965).

— — Purification and properties of an influenza virus inhibitor isolated from mouse lungs. Acta virol. **10**, 28—34 (1966).

Luzyanina, T. Ya., and R. Ya. Polyak: Localization on the termolabile virus-neutralizing factors in the protein fractions of rabbit and guinea pig sera. Yearbook Institute of Experimental Medicine of the Academy of Medical Sciences of the USSR, 1957, p. 330—332. Leningrad 1958 [in Russian].

— —, and A. A. Smorodintsev: Virus-neutralizing activity of non-specific thermolabile normal serum inhibitors against the myxovirus, adenovirus and enterovirus groups. Yearbook Institute of Experimental Medicine of the Academy of Medical Sciences of the USSR, 1960, p. 422—427. Leningrad 1961 [in Russian].

— A. L. Salminen, and A. A. Smorodintsev: Peculiarities of the interaction of thermostable inhibitors from normal sera with type A2 influenza viruses of the 1957 pandemie. Acta virol. **4**, 137—145 (1960).

—, and A. A. Smorodintsev: Neutralization of mumps virus in chick embryo cell cultures. Acta virol. **9**, 160—164 (1965).

McCrea, J. F.: Nonspecific serum inhibition of influenza haemagglutination. Aust. J. exp. Biol. med. Sci. **24**, 283—291 (1946).

— Mucins and mucoids in relation to influenza virus action. II. Isolation and characterization of the serum mucoid inhibitor of heated influenza virus. Aust. J. exp. Biol. med. Sci. **26**, 355 (1948).

Medill-Brown, M., and B. A. Briody: Mutation and selection pressure during adaptation of influenza virus to mice. Virology **1**, 301—312 (1955).

Meyer, H. M., Jr., M. R. Hilleman, M. L. Miesse, I. P. Cramford, and A. S. Bankhead: New antigenic variant in Far East influenza epidemic 1957. Proc. Soc. exp. Biol. (N.Y.) **95**, 609—616 (1957).

Morawiecki, A., and E. Lisowska: Polymerized orosomucoid an inhibitor of influenza virus hemagglutination. Biochem. biophys. Res. Commun. **18**, 606—610 (1965).

More virus inhibitors. (Leading articles.) Lancet **1961 I**, 755—756.

MULDER, J., DE NOVIJER, and L. M. BLANS: The influence of treating ferret influenza antisera with enzymes of crude filtrate of *Vibrio cholerae* on the titre of the antibodies. Antonie van Leeuwenhoek **18**, 128 (1952).
NAGASE, M., and Y. KABATA: Possible destruction of antibody by treating human serum with KIO_4. Studies on several inhibitors against influenza viruses. Virus (Jap.) **11**, 65 (1961) [in Japan with English summary].
NICULESCU, I.: Inhibitors of influenza viruses. Microbiologia, parasitologia, epidemiologia (Rumun.) **10**, 103—110 (1965) [in Rumunian].
ODIN, L.: Sialic acid in human serum protein and meconium. Acta chem. scand. **9**, 862 (1955).
ORSI, N.: Inhibitori naturali dell'infettività virale. Riv. Ist. sieroter. ital. **41**, 77—91 (1966) [in Italian].
— R. CALIO, A. STASIO, and L. SINIBALDI: On the use of influenza virus strains sensitive to non-specific inhibitors in the hemagglutination-inhibition reaction. Riv. Ist. sieroter. ital. **42**, 3—15 (1967).
PETROV, K. YU.: Effect on influenza virus of unspecific virus neutralizing substances in animal sera. Vop. Virus. **5**, 167—172 (1960) [in Russian].
PICHUSKOV, A. V.: Inhibitors of hemagglutination by parainfluenza virus types 1, 2, 3. Vop. Virus. **10**, 74—78 (1965) [in Russian].
POLYAK, R. YA., T. YA. LUZYANINA, and A. A. SMORODINTSEV: Biochemical investigation of influenza virus neutralizing protein fractions of sera from different animals. Acta virol. **3**, 61—70 (1959).
—, and A. A. SMORODINTSEV: Electrophoretic investigations of thermostable inhibitors of type A2 influenza virus in normal animal sera. Acta virol. **5**, 1—3 (1961).
RATHOVÁ, V.: The possible role of inhibitor in non-specific defence of the organism. Zprávy z epidemiol. mikrobiol. (Praha) **5**, 46—52 (1963) [in Slovak].
— (1965): Unpublished results.
— L. BORECKÝ, and D. KOČIŠKOVÁ: An attempt to influence in vivo the serum levels of myxovirus inhibitors. III. The effect of the administration of some vitamins on inhibitors and some other serum factors in guinea pigs and mice. Bratisl. lek. Listy **43**, 31—40 (1963) [in Slovak].
—, and D. KOČIŠKOVÁ: Antibody response after different modes of administration of the influenza virus group A2 in some laboratory animals and the influence of different methods of removing non-specific inhibitors on the titre of specific antibodies in their sera. Čs. Epidem. **10**, 170—175 (1961) [in Slovak].
— — J. SZÁNTÓ, O. LAUČÍKOVÁ, and L. HÁNA: Some properties of the virus-neutralizing factor of normal animal sera. Vop. Virus **4**, 717—723 (1959) [in Russian].
REIMOLD, G.: Über die Wechselwirkung von Seruminhibitoren, Hühnererythrocyten und Influenzaviren. Z. Immun.-Forsch. **118**, 121—134 (1959).
ROVNOVA, Z. I., and P. N. KOSYAKOV: Inhibitors and antibodies in the course of immunization of animals with viral preparations. Acta virol. **11**, 69—77 (1967).
RYKOWSKA, R., and L. SAWICKI: The gamma-inhibitor of normal horse serum. II. The protective properties of the serum and its fractions against infection of mice and tissue cultures with the Singapore strains of influenza virus. Med. dośw. Mikrobiol. **15**, 47—54 (1963) [in Polish].
SAMPAIO, A. A. C.: Inhibitors of influenza virus haemagglutination in normal animal sera. Bull. Wld Hlth Org. **6**, 467—472 (1952).
—, and A. ISAACS: The action of trypsin on normal serum inhibitors of influenza virus agglutination. Brit. J. exp. Path. **34**, 152—158 (1953).
SAWICKI, L., and R. RYKOWSKA: The guinea pig and white rat serum inhibitors as a possible factor of natural resistance of these animals to infection with certain strains of influenza virus. Med. dośw. Mikrobiol. **15**, 55—64 (1963) [in Polish].

Schmidt, B., K. Grossgebauer u. D. Hartmann: Über die Enzymaktivität äthergespaltener Influenzaviren und ihren Einfluß auf unspezifische Seruminhibitoren. Z. Immun.-Forsch. **120**, 238—248 (1960).

Shimizu, N., and N. Ishida: The characteristics of both hemagglutinin inhibitors and growth inhibitory substances of influenza viruses found in animal specimens. Tohoku J. exp. Med. **67**, 390 (1958).

Shimojo, H., A. Sugiura, J. Akao, and Ch. Enomoto: Studies of a non-specific hemagglutination inhibitor of influenza A (Asian) 57 virus. Bull. Inst. Publ. Health **7**, 219—224 (1958).

Shiratori, T., M. Nagase, H. Kudo, T. Kikawa, and Y. Kabata: Characterization of non-specific hemagglutinin inhibitors detectable in the sera of experimental animals. Virus (Jap.) **11**, 162—163 (1961).

Simkin, J. L., E. R. Skinner, and H. S. Seshadri: Studies on an acidic glycoprotein-containing fraction isolated from guinea-pig serum. Biochem. J. **90**, 316—330 (1964).

Smith, W., and J. C. N. Westwood: Influenza virus haemagglutination. The mechanism of the Francis phenomenon. Brit. J. exp. Path. **31**, 725—738 (1950).

— —, and G. Belyavin: Influenza: A study of four virus strains isolated in 1951. Lancet **1951 II**, 1189.

—, and M. A. Westwood: Factors involved in influenza haemagglutination reactions. Brit. J. exp. Path. **30**, 48—61 (1949).

Smorodintsev, A. A.: Factors of natural resistance and specific immunity to viruses. Virology **3**, 299—321 (1957).

—, and T. Ya. Luzyanina: Interaction of the thermolabile fractions of normal sera from various animals and the virulent and avirulent strains of the influenza virus. Yearbook Inst. of exp. Med. of the Acad. Med. Sci. USSR, 1957, p. 326—330, Leningrad 1958 [in Russian].

—, and O. I. Šiškina: The effect of normal laboratory animal sera on influenza virus. Zh. Mikrobiol. (Mosk.) **22**, 16—19 (1951) [in Russian].

Springer, G. F.: Human MN glycoproteins: Dependence of blood-group and anti-influenza virus activities on their molecular size. Biochem. biophys. Res. Commun. **28**, 510—513 (1967).

Stone, J. D.: Inhibition of influenza virus hemagglutination by mucoids. II. Differential behaviour of mucoid inhibitors with indicator viruses. Aust. J. exp. Biol. med. Sci. **27**, 557 (1949).

— Adsorptive and enzymatic behaviour of influenza virus during the O-D change. Brit. J. exp. Path. **32**, 367—376 (1951).

Stulberg, C. C., R. Schapira, A. R. Robinson, D. H. Basinski, and H. A. Freund: Inhibition of influenza virus hemagglutination by purified plasma mucoproteins. Proc. Soc. exp. Biol. (N.Y.) **76**, 704—706 (1961).

Styk, B.: Non-specific inhibitors in normal rat serum for the influenza C type virus. Fol. biol. (Praha) **1**, 207—213 (1955).

— Inhibitors of the influenza virus in human sera after passaging in volunteers. Acta virol. **1**, 238—239 (1957).

— Sensibilization by means of potassium periodate of inhibitor-resistant A2 influenza virus strains to non-specific serum inhibitors. Nature (Lond.) **193**, 954—955 (1962a).

— Effect of some inhibitor-destroying substances on the non-specific inhibitor of C influenza virus present in normal rat serum. Acta virol. **7**, 88—89 (1963).

— L. Hána, F. Franěk, F. Sokol, and J. Menšík: Investigations on cofactor and influenza antibodies by density gradient zonal centrifugation. Acta virol. **6**, 478 (1962b).

SUGIURA, A., H. SHIMOJO, and C. ENOMOTO: Studies of the non-specific hemagglutination inhibitor against influenza A2 virus (gamma-inhibitor). II. Physicochemical properties of the gamma-inhibitor. Jap. J. exp. Med. **31**, 159—168 (1961a).

— — — Studies of the non-specific hemagglutination inhibitor against influenza A2 virus. III. Destruction of gamma-inhibitor by virus enzyme. Jap. J. exp. Med. **31**, 169—177 (1961b).

SZÁNTÓ, J., O. KRIŽANOVÁ, and F. LINK: Interaction of gamma-inhibitor with A2 influenza virus. Acta virol. **6**, 524—530 (1962).

TAKÁTSY, G., and K. BARB: On the normal serum inhibitors for the avid Asian strains of influenza virus. Acta virol. **3**, (Suppl.) 71—79 (1959).

TOPCIU, V., P. DIOSI, E. MOLDOVAN, O. NEVINGLOVSCHI, and G. BRAN: Serological survey concerning the presence of influenza hemagglutination inhibitors in normal dog sera. Acta virol. **11**, 55—57 (1967).

— E. MOLDOVAN, O. NEVINGLOVSCHI, N. ROSIU, and P. DIOSI: Serological investigations on influenza inhibitors in domestic and wild animals. Rev. Roum. Inframicrobiol. **3**, 271—276 (1966).

TOUSSAINT, A. J., and L. H. MUSCHEL: Neutralization of bacteriophage by normal serum. Nature (Lond.) **183**, 1825—1827 (1959).

TYRRELL, D. A. J.: Separation of inhibitors of hemagglutination and specific antibodies for influenza viruses by starch zone electrophoresis. J. Immunol. **72**, 494—502 (1954).

VOLÁKOVÁ, N., and L. JANDÁSEK: A thermostable inhibitor of newly isolated influenza strains in guinea pig serum. Acta virol. **3**, 109—112 (1959).

VOLUYSKAYA, E. N., V. I. TOVARNICKY, and A. M. BYELIKOVA: The effect of the properdin system on the influenza virus. J. Hyg. Epidem. (Praha) **2**, 404—407 (1958).

WAGNER, R. R.: A pantropic strain of influenza virus: Generalized infection and viremia in the intact mouse. Virology **1**, 497—515 (1955).

WENNER, H. A., A. MONLEY, and M. K. JENSEN: A study of infections caused by mumps and NDV viruses. II. Some properties of specific and nonspecific serum hemagglutination inhibitor components. J. Immunol. **68**, 357—368 (1952).

WERNER, G. H., R. SHARMA, and L. GOGOLSKI: Hemagglutinin inhibition with ether-treated antigen as a more sensitive method to measure the immunological response to an Asian influenza virus vaccine. Arch. ges. Virusforsch. **10**, 7—18 (1960).

WHITEHEAD, P. H.: Viral inhibition by polymerized orosomucoid. Biochem. J. **95**, 8P (1965).

— T. H. FLEWET, J. R. FOSTER, and H. G. SAMMONS: Inhibition of influenza virus haemagglutination by polymerized orosomucoid. Nature (Lond.) **208**, No 5013, 915—916 (1965).

YAKHNO, M. A.: Influence of inhibitor-destructive treatment on influenza antibody. Vop. Virus. **6**, 334—335 (1961) [in Russian].

ZAKSTELSKAYA, L. Y., and M. A. YAKHNO: The sensitivity of different strains of A2 influenza virus to the action of inhibitors occurring in animal sera. Vop. med. Virus. (Mosk.) **5**, 71—73 (1958) [in Russian].

ZHDANOV, V. M., V. P. GAMBURG, and G. J. SVET-MOLDAVSKY: Antigenicity of the inhibitor of influenza virus strain A (Asie) 57. J. Immunol. **82**, 9—12 (1959).

— L. J. ZAKSTELSKAYA, V. E. JEFIMOVA, M. A. YAKHNO, and S. L. KHAIT: Antigenic peculiarities of 1957 influenza viruses and serologic indexes of immunity among the population. J. Amer. med. Ass. **167**, 1469—1474 (1958).

Biochemical Research Laboratory, Massachusetts General Hospital, Harvard Medical School, Boston, Mass./USA

Cellular Membrane Alterations in Neoplasia: A Review and a Unifying Hypothesis

DONALD F. HOELZL WALLACH

Table of Contents

A. Introduction

Together with its abnormal growth, the neoplastic state is characterized by numerous and variable cellular derangements, but none of the many biochemical, immunological, biological and morphological defects hitherto described is common to all tumors. Moreover, even tumors induced by a single agent in a single cell type can have a multiplicity of phenotypes, which tend to modulate in time. Pleiomorphism thus constitutes a dominant aspect of neoplasia.

Current evidence suggests that there may be a common feature underlying the diversity of neoplasia; namely, that alterations in one or more of their membrane systems is a regular feature of tumor cells. In the following I will review both the direct evidence for membrane defects and data which implicate various membrane functions, although circuitously. I will then submit a hypothesis, consistent with the many diverse observations and accounting for the individuality and pleiomorphism typical of tumors.

B. Plasma Membrane

The plasma membrane is the site of many alterations fundamental to the essential biologic process of "malignancy" — invasiveness and metastasis, COMAN (1944, 1953); HUXLEY (1958), — and to the immunology of tumors. However, no single feature is an invariant aspect of neoplasia.

I. Cell Contact

1. Cellular Adhesiveness

The social relationships of neoplastic cells with each other and with normal cells are defective, HUXLEY (1958), ABERCROMBIE (1962), CURTIS (1967), STOKER (1967a), a fact which COMAN (1944, 1953, 1954, 1960, 1961), COMAN and ANDERSON (1955), interpreted in terms of decreased "mutual adhesiveness" of neoplastic cells, COMAN and ANDERSON (1955), CURTIS (1960). However, low "mutual adhesiveness" is not a necessary aspect of "malignancy". Thus HALPERN *et al.* (1966) report the malignant variants of several tissue culture lines to be more adhesive when grown in agitated media than their normal progenitors. This is in contrast to the common experience that malignant cells grown on solid media adhere less to each other than do normal cells. Whatever the explanation, this and other observations indicate that cellular adhesiveness changes during neoplastic conversion and that both increased and decreased "mutual adhesiveness" can occur.

2. Contact Inhibition of Movement

This well known effect was first described by ABERCROMBIE and associates (1954, 1958), who showed by time lapse cinematography of cells migrating from the chick embryo and other explants that contact of the active "ruffled" border of mobile cells with the surface of other cells led to the immobilization of the cell membranes, causing cessation of the movement of the cells towards or over each other. However, the "non-contacted" surfaces of the cells would continue to ruffle and probe and the cells would continue to move on the free area of the solid support until a solid monolayer of contiguous and immobile cells had formed.

As first reported by ABERCROMBIE *et al.* (1957) tumor cells commonly lack contact inhibition of movement. This defect appears very early in the neoplastic conversion of various cells by oncogenic viruses, TEMIN and RUBIN (1958); VOGT and DULBECCO (1960), SACHS and MEDINA (1961), MACPHERSON and STOKER (1962), STOKER (1964). However, more extensive study of the phenomenon has shown that some tumor cells which are not inhibited by association with like cells, are inhibited by contact with normal cells or tumor cells of different origin, STOKER (1964), BARSKI and BELEHRADEK, JR. (1965), BOREK and SACHS (1966), STOKER *et al.* (1966). Thus, BOREK and SACHS (1966) find that fibroblasts converted by polyoma virus may inhibit and be inhibited by SV_{40}-converted cells but not by each other and that doubly converted cells can be inhibited by singly converted cells of either type. This work deserves amplification in that it suggests that contact inhibition is highly specific.

3. Contact Inhibition of Growth

Another important aspect of cell contact is the fact that, as a contact inhibited layer approaches confluency, there is a rapid shutoff of net RNA and protein synthesis and cessation of DNA synthesis, STOKER (1967a), LEVINE *et al.* (1965). In mouse 3T3 lines, contact inhibition of growth occurs in the postmitotic stage (G1), TODARO *et al.* (1964, 1965, 1966), NILAUSEN and GREEN (1965); and TODARO *et al.* (1966) argue that certain macromolecules in the cell environment act as intermediaries in the process. These substances may be related to the growth-promoting material found by RUBIN (1966) in the medium of dense populations of chick embryo cells. Contact inhibition of grow this typically much less prominent in neopastic cells than in normal cells, STOKER (1967a).

4. Electrical Coupling and Molecular Transfer

It has been well established that electrical connections — via ionic exchange — can occur between individual cells of various epithelia, POTTER *et al.* (1966); LOEWENSTEIN *et al.* (1965); LOEWENSTEIN (1966). These ionic connections depend upon the presence of Ca^{2+} or Mg^{2+} in the medium, NAKAS *et al.* (1966). There are extensive electrical connections between the hepatocytes in normal or regenerating rat liver, PENN (1966); LOEWENSTEIN and PENN (1967), but LOEWENSTEIN and KANNO (1967) could find no coupling between neoplastic hepatocytes in a number of hepatomas nor between malignant thyroid or gastric cells, JAMAKOSMANOVIC and LOEWENSTEIN (1968); KANNO and MATSUI (1968). However, there is good coupling between normal and neoplastic fibroblasts in culture, POTTER *et al.* (1966). Indeed, FURSHPAN and POTTER (1968; personal communication), find identical electric coupling — and fluorescein transfer — between BHK21 or 3T3 cells before or after conversion by polyoma or SV_{40}, i.e., even between cells lacking contact inhibition. In the case of Sarcoma 180 cells in tissue culture, coupling is good about 12 hours after replating, then diminishes, only to rise again as the cultures become dense.

The above described cellular communications can act as an important mechanism of intercellular metabolic control. This is apparent from the studies of SUBAK-SHARPE (1965) and STOKER (1967b), demonstrating alleviation of defective nucleic acid metabolism in a mutant of polyoma-converted BHK21 cells by transfer of an undefined substance from wild type cells during cell contact. It remains to be seen whether the defective contact inhibition of neoplasia is due to aberrant intercellular transfer of regulatory substances or to abnormal contact.

5. Cell Fusion

The limit of cell contact is cell-fusion and this can be brought about *in vitro* by certain myxoviruses. The mechanism involved in this process have been investigated in detail by OKADA and associates, OKADA (1962); OKADA and TADOKARO (1962); OKADA *et al.* (1966); OKADA and MURAYAMA (1966), mostly using Ehrlich ascites carcinoma. From a study comparing a number of diverse cell types, OKADA and TADOKORO (1963) suggest that their "fusion capacity" parallels their malignancy.

II. Surface Charge

1. Electrophoresis

The interaction between individual cells has long been considered to depend on the charge density on cell surfaces. This notion arises from the fact that small particles, dispersed in aqueous media and subject to random thermal motion, tend to stick upon collision, unless electrostatic repulsions between their surfaces are great enough to make the collisions elastic, WALLACH *et al.* (1966). Therefore, many investigators have attempted to explain the altered contact behavior of neoplastic cells by abnormalities of cell surface charge. Unfortunately, there has been no study correlating charge density, adhesiveness and contact inhibition.

However, several internally controlled experiments show changes in the electrophoretic mobility of cell populations during neoplastic conversion. Thus, hamster kidney cells have a lower electrophoretic mobility than cells from stilbesterol-induced kidney tumors, AMBROSE *et al.* (1956), and the anodic mobility of hepatoma cells is greater than that of normal hepatocytes, LOWICK *et al.* (1961). Also PURDOM *et al.* (1958) show the surface potential of mouse MCIM sarcoma sublines to increase as these progress from the solid to the ascites form, a change associated with greater invasiveness, and FORESTER *et al.* report an increase in the anodic mobility of hamster fibroblasts upon conversion with polyma virus, FORESTER *et al.* (1952, 1964) and of mouse spleen cells in Friend virus disease, FORESTER and SALAMANN (1967). Finally, certain leukemic mouse cells have an abnormally *low* anodic mobility, COOK and JACOBSON (1968).

These results appear at variance with the seeming lack of correlation between anodic mobility and malignancy in a number of human, VASSAR (1963), and mouse, SIMON-REUSS *et al.* (1964), tumors, but the latter studies are technically weak in that they do not compare the electrophoretic mobility of a certain cell type before and after neoplastic conversion, using identical experimental conditions. There are too few such controlled studies to allow generalizations, but it does appear that neoplastic conversion can change the cell surface potential in some instances, FORESTER *et al.* (1952, 1964); FORESTER and SALAMANN (1967); COOK and JACOBSON (1968).

2. Specific Ionogenic Groups

A major contribution to the negative surface potential of mammalian cells is sialic acid (N-acetyl-neuraminic acid), CURTIS (1967), but charged side chains of membrane proteins, WALLACH *et al.* (1966); COOK and JACOBSON (1968); WALLACH and KAMAT (1966), membrane RNA, WEISS and MAYHEW (1966), and presumably membrane lipids also contribute to the net cell surface charge.

Several authors have proposed that the possible increase in negative surface charge on neoplastic cells is due to increased surface sialic acid, but there is at present insufficient evidence for this generalization. Indeed OHTA *et al.* (1968) found three lines of virus-converted fibroblasts (including polyoma-converted BHK21 cells) to have less total sialic acid than their normal precursors. This is surprising since polyoma-conversion of BHK21 cells causes a sialic acid-linked rise in electrophoretic mobility, FORESTER *et al.* (1952, 1964) but OHTA *et al.* measured only total sialic acid and may have missed opposite changes in membrane sialic acid. However, WU and associates (1968) report that there really is a decrease in the sialic acid (and galactosamine) of the plasma membrane and endoplasmic reticulum of 3T3 cells converted by SV_{40} virus. But even these studies, showing an alteration of membrane sialic acid in neoplastic conversion, cannot readily be related to surface potential, because the electrokinetic expression of ionic groups on membranes depends, among more complex variables, upon the effective radius of the charge-bearing site, on other charges and upon the depth of the charge in the membrane, all of which are unknown, WALLACH *et al.* (1966); WALLACH and KAMAT (1966); WALLACH and PEREZ-ESANDI (1964). This point is illustrated by the low anodic mobility of certain leukemic cells of mice which actually have more sialic acid on their surfaces than normal, COOK and JACOBSON (1968). In this case the increased charge contribution of sialic acid is outweighed by a concomitant rise in surface cationic groups, COOK and JACOBSON (1968).

There are thus at least four instances — the conversion of BHK21 cells by polyoma virus, FORESTER *et al.* (1952, 1964), Friend virus disease, FORESTER and SALAMANN (1967) and certain leukemias, COOK and JACOBSON (1968) — where the abnormal anodic mobility of the converted cells is modified by enzymatic (neuraminidase) removal of sialic acid. The possible interrelations between surface charge, sialic acid and/or other ionogenic groups and malignant behavior thus deserves greater study.

III. "Leakiness" of the Plasma Membrane

The plasma membranes of tumor cells commonly exhibit abnormal "permeability" to certain intracellular enzymes, HILL (1956); MCNAIR-SCOTT *et al.* (1959); WROBLEWSKI (1958); WU (1959); WU and RACKER (1958); HOLMBERG (1961); MALMGREN *et al.* (1955); SYLVEN (1958, 1962); SYLVEN and BOIS (1960); SYLVEN and MALMGREN (1955, 1957); SYLVEN *et al.* (1959). The preferentially released enzymes are lysosomal and/or those usually considered free in the cytoplasm. The "leakiness" observed is not due to cell lysis. Highly "leaky" cells are normal by the usual criteria of dye exclusion and various students of the phenomenon attribute it to a peculiarity of the plasma and/or lysosomal membranes of tumor cells.

The matter has been studied with great care by SYLVEN and associates (1959) with emphasis on possible correlations to invasiveness. Using carefully controlled and highly elegant micropuncture and microanalytical procedures, these workers demonstrated the release of lysosomal peptidases and hydrolases from intact tumor cells into the interstitial fluid surrounding them, particularly at the invasive zone. They suggest that these enzymes participate in the destructive behavior of invasive tumors and conjecture that enzymatic attacks of the tumor cells *per se* might be involved in their decreased "mutual adhesiveness".

The mechanisms of translocation of macromulecules across plasma membranes of normal or neoplastic cells remain very poorly understood, RYSER (1968). It is also not known whether abnormal plasma membrane "leakiness" is a general property of neoplasms, or whether it is a manifestation of an immune reaction against the tumor, but it is of considerable importance to our understanding of the basic defects in neoplasia and of the mechanisms of invasiveness. Study of the phenomenon in cells transformed by oncogenic viruses and also in minimal deviation hepatomas should prove particularly valuable.

IV. Immunologic Changes

1. New Antigens

Those plasma membrane alterations of neoplastic conversion, which are the best defined genetically, are immunologic ones. Plasma membranes of tumors, induced by chemical carcinogens and by viruses, bear transplantation antigens not present in the tissues of origin, ALEXANDER (1966); HELLSTROM (1965); OLD and BOYSE (1964); SJOGREN (1965); KLEIN (1967 1968); HARRIS (1967). It is not clear at present whether the new antigens are essential to the cancerous state, but they play a critical role in the tumor-host relationship, since they can effect the immunological elimination of neoplastic cells under favorable circumstances. The new antigen of any one, *chemically*-induced tumor, appears to be specific to that tumor, but tumors induced by a given *virus* always bear the same new transplantation antigen, regardless of the tissue or species of origin. One cannot say at present whether the new antigens are on the plasma membrane exclusively.

2. Embryonic Antigens

The immunologic anomalies of tumor plasma membranes are not limited to the appearance of entirely new antigenic determinants. Thus, the neoplastic transformation of hamster fibroblasts by polyoma and other viruses leads to the appearance of Forssman antigen on the surface of the transformed cells, O'NEILL (1968); VOGEL and SACHS (1962, 1964). Forssman antigen is lacking in fibroblasts of neonatal and mature hamsters but is present in embryonic hamster cells. The appearance of embryonic antigens has also been shown to occur in several human tumors, GOLD and FREEDMAN (1965a, 1965b); GOLD *et al.* (1968). The nature of Forssman determinants is quite well understood, MAKITA and YAMAKAWA (1962); MAKITA *et al.* (1966), and glycolipids bearing these determinants have been isolated in pure form from many of Forssman-positive tissues.

Further evidence for specific antigenic alterations of the plasma membrane during neoplastic transformation comes from the observation that certain neoplastic cells, but not their normal progenitors, are agglutinated by a phytoagglutinin from wheat germ, AUB *et al.* (1963); BURGER and GOLDBERG (1967), a fact which HAKOMORI and MURAKAMI (1968) trace to the altered glycolipid composition of neoplastic plasma membranes. They report a four-fold drop of cellular hematoside and a ten-fold rise of lactosylceramide as a result of the conversion of BHK21 cells by polyoma virus and provide data suggesting that the carbohydrate moiety of the neoplastic glycolipids is incomplete. Relating their observations to the deletion of blood-group A and B haptens in some human adenocarcinomas (with simultaneous accumulation of Le^A and H-glycolipids), HAKOMORI *et al.* (1967); MASAMUNE *et al.* (1952), they argue for abnormal synthesis of carbohydrate chains in tumors, due to a defect in/or inhibition of enzymes participating in carbohydrate metabolism.

3. Antigen Deletion

Neoplastic conversion may also produce deletion of certain organ specific antigens. This has been studied extensively in hepatocellular carcinomas and has been recently reviewed by ABELEV (1965). Hepatomas usually lack some but not all of the liver-specific antigens and there is typically considerable variation in antigenic composition from one newly induced hepatoma to another. However, after many transplant generations, diverse tumors tend to arrive at a common antigenic pattern.

C. The Mitochondrial Membrane

I. Abnormal Regulation of Glycolysis — "Warburg Phenomenon"

Most normal tissues sharply reduce glycolysis when there is sufficient oxygen available for oxidative phosphorylation (Pasteur effect). This regulatory mechanism is commonly defective in tumors, WENNER (1967); MORRIS (1965). Extensive efforts to implicate some one glycolytic enzyme in the "Warburg effect" have not yielded definitive results, probably because almost every step in glycolysis is under some sort of direct or indirect control through intermediates which also participate in mitochondrial metabolism, ATKINSON (1965); CHANCE and HESS (1959); RACKER (1965); WOOD (1966). It now appears that the "Warburg effect" is due to mitochondrial malfunction.

There are two feedback systems linking glycolysis with oxidative phosphorylation. Both involve obligatory intermediates, NAD/NADH and ADP/ATP, neither of which can freely permeate the mitochondrial membrane. The translocation of the reducing equivalents of NADH from cytoplasm to mitochondria is commonly abnormal in tumors, BOXER and DEVLIN (1961). In many instances this involves impaired activity of the acetoacetate-β-hydroxybutyrate "shuttle" of the mitochondrial membranes, BOXER and DEVLIN (1961). However, other shuttle systems, not involving known "membrane enzymes" are also frequently defective.

Other recent studies, ELLSWORTH *et al.* (1963); LIN *et al.* (1962); LO *et al.* (1968); WEBER *et al.* (1961) also strongly implicate the mitochondria in the "Warburg phenomenon" and have been interpreted to support the suggestion by JOHNSON (1941) and LYNEN (1941) that the Pasteur effect is due to competition for ADP and P_i by the phosphorylation sites of glycolysis and respiration. They find that "minimum deviation" hepatomas fall into two main classes: well differentiated, slowly growing tumors with low aerobic glycolysis and moderate to high respiration, and poorly differentiated, rapidly growing tumors with high aerobic glycolysis and low respiration. The latter contain only 9 to 14 mg of mitochondrial protein per gram of tissue, whereas the range for well-differentiated tumors is 19 to 33 mg and for normal liver about 50 mg, OHE *et al.* (1967). The low mitochondrial content and high levels of glycolytic enzymes of dedifferentiated tumors provide one possible basis for the "Warburg phenomenon". However, the recent report of LO *et al.* (1968) suggests that another mechanism is also operative. In these studies the interaction between the mitochondria and glycolytic enzymes from liver and various tumors were investigated. The critical finding is that the high glycolysis in homogenates of many tumors can be abolished by replacing their mitochondria with those from tissues having a normal Pasteur effect. Since ADP levels control oxidative as well as glycolytic phosphorylation, the defective control of glycolysis by the mitochondria of many tumors could be related to an abnormal distribution of this intermediate among the compartments of the tumor cell. Neither ADP nor ATP can cross the inner mitochondrial membrane by diffusion, BRIERLY and GREEN (1965); KLINGENBERG and PFAFF (1966);

rather, permeation appears to involve specific enzyme systems located in the inner mitochondrial membrane. BRIERLY and GREEN (1965) have argued for *transphosphorylation* systems in the mitochondrial membranes, but the elegant work of KLINGENBERG and PFAFF (1966) points to *translocation* of ATP and ADP (not AMP, GDP, NAD, NADP, etc.) by specific enzyme systems located in the inner mitochondrial membrane. The studies of NEIFAKH and co-workers (1961) suggest that the permeability of mitochondria undergoes reversible dynamic changes depending on ATP levels within the mitochondria. Thus, GAITSKHOKI (1961; 1962) reports that the rate of glycolysis in liver is limited by low concentration of adenine nucleotides and that the permeability of the mitochondria to these substances is significant in regulating glycolysis. He argues that this movement of co-enzymes depends on reversible changes in the structure of the mitochondrial membrane and that this control is defective in tumor cells. This suggestion is certainly consonant with other evidence indicating that high aerobic glycolysis, the most widely reported metabolic aberration of tumors, relates to defective membrane-mediated control mechanisms.

II. Aberrant Regulation of Fatty Acid Biosynthesis

The mitochondria may also be involved in the defective regulation of *de novo* fatty acid biosynthesis observed in all hepatomas tested to date, SABINE *et al.* (1966, 1967, 1968); MAJERUS *et al.* (1968), although this control is functioning normally in the host livers. The defect is present even in highly differentiated, slowly growing hepatomas. Tumors appear well equipped to utilize exogenous fat, SPECTOR (1967); SPECTOR and STEINBERG (1967). The rate-determining step in normal synthesis is the ATP-requiring formation of malonyl-CoA from acetyl-CoA and CO_2, catalyzed by cytoplasmic acetyl-CoA-carboxylase, NUMA *et al.* (1965); SHAPIRO (1967); FRITZ (1967); FRITZ and HSU (1967). The activity of this biotin-enzyme in livers of animals on normal diet is 1/10th that of the fatty-acid synthetase complex, which converts malonyl-CoA to palmityl-CoA. Fat feeding or fasting produces a further decrease of the already limiting rate of acetyl-CoA-carboxylation and thus depression of *de novo* fatty acid synthesis. *In vitro* studies indicate that acetyl-CoA-carboxylase is an allosteric enzyme which is reversibly activated by citrate and reversibly inhibited by long chain fatty acids and their CoA derivatives. The inhibition produced by these long chain compounds *in vitro* as well as the dietary inhibition of acetyl-CoA-carboxylase can be rapidly and fully reversed by another fatty acid derivative — fatty acyl-carnitine, FRITZ (1967); FRITZ and HSU (1967). The inhibition of the carboxylation reaction is an example of end-product inhibition of a rate-controlling step, but the situation is unique in that the inhibition can be turned off by conversion of the inhibitor to a related intermediate, fatty acyl-CoA, by a transferase located on the inner mitochondrial membrane, FRITZ (1967); FRITZ and HSU (1967). There it mediates, in a shuttle mechanism, the permeation of fatty acids into the mitochondrial space, and thus their catabolism. In thus stimulating the removal of fatty acids from the cytoplasm, fatty acyl-carnitine indirectly controls fatty acid synthesis. It also reverses the inhibition of acetyl-CoA carboxylase by fatty acids through direct interaction with the enzyme, FRITZ (1967); FRITZ and HSU (1967).

The acetyl-CoA-carboxylase purified from several hepatomas appears to be identical to that of the host livers, and the loss of regulation of this enzyme activity in hepatomas has been attributed to increased amounts of the carboxylase, MAJERUS *et al.* (1968). But the problem can profitably be viewed differently, i.e., as one of membrane-mediated control. Thus, regulation of *de novo* fatty acid synthesis will depend on the steady state concentrations of fatty acyl CoA and of fatty acyl-carnitine. The former are in turn determined by the permeation of fats and fatty acids across the plasma membrane, their activation and conversion to glycerides in the endoplasmic reticulum, and their permeation into and oxidation by the mitochondria. The level of the carnitine derivative is controlled by mitochondrial acyl-CoA-carnitine transferase. It would clearly be fruitful to test whether the defect seen in hepatomas does not arise from 1) abnormal permeability of the mitochondria to the inhibitor (fatty acyl-CoA); 2) unusually effective conversion of the inhibitor to fatty acyl-carnitine by the tumor mitochondria; or 3) abnormal fatty acid metabolism at sites in the plasma membrane or endoplasmic reticulum.

III. Mitochondrial Protein Synthesis

A preliminary report by GRAFFI and associates (1965), on the difference between mitochondrial protein synthesis in normal and neoplastic cells is highly pertinent to the question of possible mitochondrial abnormalities in neoplasia. Studying the *in vitro* incorporation of leucine and arginine (guanidine label) into mitochondrial proteins, these authors find that in many tumors the newly synthesized mitochondrial protein has a leucine/arginine molar ratio of 0.05 to 0.27, whereas the mitochondria of embryonic and adult tissues had ratios of 2.0 to 3.0. Only the mitochondria from mouse, rat and bovine kidney and bovine liver have low ratios (0.0006 to 0.05), which are attributed to urea metabolism. The protein synthesized by mitochondria is primarily "structural protein" and the above results are interpreted to indicate synthesis of an abnormal "structural protein" in neoplasia. Before the significance of this finding can be fully evaluated, studies on more comparable normal and neoplastic cells are needed.

D. Lysosomes

There is little specific information on the membranes of tumor lysosomes, but the high activities of lysosomal enzymes in the interstitial fluids surrounding healthy tumor cells *in vivo* implicate both the surface and lysosomal membranes, SYLVEN (1962).

E. The Nuclear Membrane

The nuclear membrane is not well understood, but recent measurements of electrical impedance, WIENER *et al.* (1965), show that it is not a porous structure and that it might therefore regulate the movement of various controlling substances, and also of mRNA between nucleus and cytoplasm. There is no specific information about the nuclear membrane of tumors, but recent information relating membranes to the initiation of DNA replication in microbial, JACOB *et al.* (1963); GANESAN and LEDERBERG (1965); RYTER and JACOB (1963); C. LARK and G. LARK (1964); LARK (1966); JACOB *et al.* (1966); RYTER (1968), and animal, COMINGS and KAKEFUDA (1968), cells may bear on the nuclear aberrations commonly seen in tumors. There appears to be no definite correlation between growth rate and/or biochemical deviation and nuclear morphology and/or chromosome number, WU (1967).

F. Endoplasmic Reticulum

The endoplasmic reticulum (E.R.) of neoplastic cells is markedly different from that of normal cells and this fact has produced some stimulating speculations on molecular mechanisms of neoplasia and carcinogenesis, PITOT (1964, 1966); PITOT and CHO (1965). The most extensive information concerning E.R. changes in neoplasia comes from studies of hepatomas, but there is a recent report on a change of E.R. carbohydrates as a consequence of the *in vitro* transformation of mouse fibroblasts by SV_{40} virus, WU *et al.* (1968).

I. Morphologic Changes

The earliest morphologic aberrations during hepatic carcinogenesis are disorganization of the E.R., a disappearance of rough E.R., appearance of free polysomes and a loss of glycogen granules; but the ultrastructure depends upon growth rate, and slowly growing tumors are the least deviant morphologically, DALTON (1964).

II. Polysome Binding

Hepatomas contain a smaller proportion of heavy polysomes and a higher proportion of ribosome monomers and dimers than normal liver, WEBB *et al.* (1964, 1965); the extent of the anomaly correlates roughly with the degree of metabolic deviation of the tumor. Moreover, the fraction of ribosomes which are membrane-bound is 60 to 70% in normal and regenerating adult liver, but 20% or less in well differentiated hepatomas and 0 in highly deviant hepatomas, WEBB *et al.* (1964, 1965). The defective binding of polysomes to hepatoma E.R. appears to be a function of the membranes and not the polysomes, SUSS *et al.* (1966). It is of interest in this context that in the malignant HeLa cell only 10 to 15 % of the polysomes are membrane-bound and these bear a mRNA, which from its sedimentation characteristics, base composition, sequence homology with cytoplasmic DNA and metabolic behavior is distinct from the mRNA of free polysomes, and is presumed to be synthesized on a mitochondrial DNA template, B. ATTARDI and G. ATTARDI (1967).

III. Enzymatic Changes

Associated with the morphologic changes are extensive alterations in the enzymes of the "microsomal fraction" (which contains endoplasmic reticulum and plasma membrane, WALLACH (1967)), but the enzymatic deviations which occur during hepatic carcinogenesis do not correlate well with either growth rate or nuclear morphology, WU (1967). Indeed, as is true immunologically, each chemically-induced tumor is an entity unto itself; a rapidly growing tumor with highly deviant nuclear morphology may be less enzymatically deviant than more differentiated tumors. To date only two microsomal enzyme systems have been shown to be altered in all hepatomas tested, namely Mg^{2+}-ATPase, which is always elevated, SUGIMURA *et al.* (1966), and the negative feedback system regulating cholesterol synthesis, which is always deficient, SIPERSTEIN and FAGAN (1964); SIPERSTEIN *et al.* (1966).

Microsomal enzymes vary widely from one tumor to the next and their activities are generally at equivalent or somewhat lower levels in hepatomas than in normal and regenerating liver, SUGIMURA (1966). This irregular phenotypic expression extends also to the "inducible" drug-metabolizing enzymes of "smooth" microsomal membranes, HART *et al.* (1965). Both minimal deviation hepatomas and more deviant tumors are devoid of or deficient in at least one of the drug-metabolizing enzymes of the smooth microsomal membranes. The minimum deviation hepatomas had some activity for most of the enzymes and this increased upon administration of phenobarbital. However, all minimum deviation hepatomas have at least one defect. No basal enzymes could be detected in the Novikoff hepatoma but some were inducible with phenobarbital.

The studies of hepatoma microsomes have revealed not only abnormal levels and abnormal inducibilities of membrane-associated enzymes, but also an instance of defective feedback control. The rate of cholesterol synthesis from two carbon precursors by normal liver can be greatly depressed by cholesterol in the diet, the feedback mechanism residing in an early step in the biosynthesis of cholesterol, namely, at the conversion of beta-hydroxyl-beta-methylglutaryl-CoA to mevalonic acid. This reaction is catalyzed by a reductase which requires NADPH and CoA and is found primarily in the microsomal membranes of normal, TOMKINS *et al.* (1953); SIPERSTEIN *et al.* (1960); SIPERSTEIN and FAGAN (1966), and neoplastic, SIPERSTEIN and FAGAN (1964); SIPERSTEIN *et al.* (1966), liver. Some activity is also found in the soluble fraction but this is not subject to feedback inhibition. The feedback control of cholesterol biosynthesis is lacking in all hepatomas (mouse, rat, human) studied, regardless of rates of lipo- or cholesterogenesis, even though functioning well in the liver of the tumor-bearing animal and regenerating liver, SIPERSTEIN and FAGAN (1964); SIPERSTEIN *et al.* (1966).

IV. Altered Regulation of Enzyme Biosynthesis

PITOT and his associates, (1964, 1965, 1966) have studied in detail, the control of biosynthesis of several amino-acid catabolizing enzymes in hepatomas and normal liver. Their findings fit the generalization that each hepatoma has its own enzymatic phenotype, but they also show that the observed alterations are not due to changes in the enzymes *per se*, but rather to altered control of their biosynthesis. They present evidence that the altered control is due to differences in stability of specific mRNAs between normal liver and tumors and between one tumor and the next. They argue from observations on other biologic systems, ARONSON (1965); PITOT *et al.* (1965); LAMAR *et al.* (1966), that mRNA stabilization is a membrane function, that the E.R. has this function in animal cells and that neoplastic alterations of E.R. membranes can lead to increased or decreased stability of a given mRNA.

G. Discussion

Although much of the evidence cited above was collected to test other hypotheses, it does suggest that neoplasms typically have defects in one or more cellular membranes. But does this realization aid our fundamental understanding of tumor biology? Can a unitary membrane defect be postulated which accounts for the pleiomorphism and variability characteristic of tumor behavior?

Our conception of cellular membranes is at present in a state of flux, particularly since traditional, lipid-based theories of membrane structure, DANIELLI and DAVSON (1935); ROBERTSON (1959), do not explain the multiplicity of functions associated with cellular membranes, KORN (1966); MADDY (1966); ROSENBERG (1967). However, recent studies on diverse cellular membranes by freeze-cleaving or freeze-etching electron microscopy, MOOR (1966); WEINSTEIN and WILLIAMS (1967); WEINSTEIN and KOO (1968), and by more conventional electron microscopy, SJOSTRAND (1968), suggest that these structures are arrays of globular lipoproteins. Also recent investigations of various cellular membranes by optical rotatory dispersion and circular dichroism, WALLACH and ZAHLER (1966); LENARD and SINGER (1966, 1968); WALLACH and GORDON (1967, 1968a, 1968b); URRY *et al.* (1967); STEIM (1967), indicate that membrane proteins have a considerable proportion of their peptide bonds in the alpha-helical conformation, and infrared spectroscopy, WALLACH and ZAHLER (1966, 1968c); STEIM (1967); MADDY and MALCOLM (1965; 1966); CHAPMAN *et al.* (1968) indicates that plasma membranes lack the β-conformation envisaged in traditional membrane models. The optical activity measurements of WALLACH and associates (1966, 1968a, 1967, 1968b), of LENARD and SINGER (1966, 1968) and of URRY *et al.* (1967) also suggest close hydrophobic associations between the helical segments of membrane proteins and/or between helical segments and hydrocarbon portions of membrane lipids. Hydrophobic association between membrane proteins and lipids has been stressed by GREEN and FLEISCHER (1963) on chemical grounds but more direct evidence for it comes from the proton magnetic resonance spectra, CHAPMAN *et al.* (1968), and infrared spectra of cellular membranes, MADDY and MALCOLM (1965, 1966); WALLACH and ZAHLER (1968); CHAPMAN *et al.* (1968).

Recent data thus suggest that large regions of cellular membranes are *lattices* of *interacting lipoproteins*, WALLACH and GORDON (1967, 1968b); CHANGEUX *et al.* (1967); CHANGEUX and THIERY (1967); CHANGEUX and PODLESKI (1968); CASPAR (1968); GREEN and PERDUE (1966); LEHNINGER (1968); WATANABE *et al.* (1967); NACHMANSOHN (1966). In such a system any one subunit or protomer, CHANGEUX and THIERY (1967), would of necessity modulate the behavior of other members of the lattice. Moreover, heritable changes of one type of protomer could alter not only its own specific properties but also those of other members of the lattice. This is implicit in the mathematical analysis of CHANGEUX *et al.* (1967) of cooperativity in membranes. These authors point out that membrane changes may become *permanent* as a result of the insertion of protomers of a different structure or unit size.

If all cellular membranes have certain "protomers" in common, permanent alterations of such proteins would have far-reaching effects on the functions associated with all the various membrane systems. This view derives from the recognition, originating in the study of hemoglobin, PERUTZ (1965); PERUTZ *et al.* (1963), that the function, and even specificity, of soluble enzymes is determined not only by themselves, but also by their macromolecular associates. The pertinence of these notions to membranes is brought out strikingly in the finding of GANSCHOW and PAIGEN (1967) that the *presence of enzyme* activity in a membrane can be determined by factors other than its primary structure and that separate genes determine these factors. They show that certain strains of mice lack glucuronidase in their hepatic endoplasmic reticulum (but not lysosomes) due to mutation of a Mendelian gene, distinct from the glucuronidase structural gene.

Closely related is the work of RAUCH (1968) demonstrating multiple, membrane-associated defects in the "dilute lethal" mutation of mice.

To account for the pleiomorphism of tumors by a membrane hypothesis requires that various cellular membranes share important protein constituents. HULTIN *et al.* (1964) were first to suggest that a class of closely related "structural" proteins is common to membranes of diverse sources. Also, WALLACH and VLAHOVIC (1967) showed immunologically that many of the proteins in the plasma membrane of Ehrlich ascites carcinoma are also present in endoplasmic reticulum. Finally, the recent studies of WOODWARD and MUNKRES (1966a, 1966b, 1967, 1968) show that a protein similar in physical characteristics, peptide mapping, and amino acid composition can be isolated from the plasma, nuclear, microsomal, and mitochondrial membranes of *Neurospora* and various *Neurospora* mutants by procedures similar to those used for the isolation of "structural" proteins from animal mitochondria. Like its counterpart in animal cells, this protein has the unique ability to complex a variety of mitochondrial enzymes, phosphatides and nucleotides.

Most important are the studies of WOODWARD and MUNKRES (1966a, 1966b, 1967, 1968) on several mutants of *Neurospora* in which multiple mitochondrial or membrane functions appear either in excess or deficit. They show that the pleiomorphic defects of these mutants arise from single amino acid replacements in the membrane "structural" protein. For example, in the mutant mi-1, a tryptophan is exchanged for a cysteine and another mutant protein, (mi-3), also lacks one of the three tryptophan residues, but does not have an extra cysteine. In both cases only mutant "structural" protein was found in the various cellular membranes. The presence of the abnormal "structural" protein in the plasma membrane is thought to be related to the altered colony morphology of the mutant cells.

Compared with the mitochondria of wild-type *Neurospora* those of the mutant mi-1 have a marked deficit in cytochrome b, cytochrome a and a_3, cytochrome oxidase and succinic acid dehydrogenase and catalase, M. MITCHELL and H. MITCHELL (1952); HASKINS *et al.* (1953); TISSIERES *et al.* (1953); TISSIERES and MITCHELL (1954). In contrast cytochrome c, TISSIERES and MITCHELL (1954), and unsaturated fatty acids, HARDESTY and MITCHELL (1963), are present in large excess in the mutant mitochondria.

The mutant "structural" proteins are strikingly abnormal in their interactions with ATP, NADH and malic dehydrogenase. Thus, mi-1 "structural" protein binds ATP, NADH and malic dehydrogenase much more strongly than the wild-type "structural" protein. Similarly mi-3 has a higher affinity for ATP than normal. Moreover, when complexed with mutant "structural" proteins, wild-type malic dehydrogenase has a K_M which is very much larger than that of the enzyme complexed with wild-type "structural" protein, the enzyme in wild-type intact mitochondria or the free enzyme. In contrast, the combination of wild-type fumarase with mutant "structural" protein changes its V_{max} but not its K_M, WOODWARD (1968).

The pleiotropic effects of membrane-associated mutations are not seen only in Neurospora. Thus, TUPPY, SWETLEY and WOLFF (1968) relate the pleiomorphic alterations of the respiration-deficient "petite" mutant of *saccharomyces cervisiae* (D-273-10B-1) to loss of a mitochondrial "structural" protein component. Further, AZOULAY *et al.* (1967) have isolated mutants of *E. coli* K12, characterized by the simultaneous loss of several enzymatic activities, all of which are membrane-associated.

The special effects of the mutant membrane proteins described above lead one to suspect the existence of a whole spectrum of membrane mutants[1] with widely varying properties, and one can anticipate that such mutants will be recognized with increasing frequency as adequate methods for the fractionation of membrane proteins become available. The new properties of membrane mutants might arise simply from the altered function of the mutant protein, or they may be pleiomorphic due to the influence of the altered protein on normal members of the membrane lattice.

Within the class of membrane mutants only those in which the *plasma membrane* is critically involved would appear as neoplasms. This is because neoplasia is a disorder in the social interaction between cells which evolves at their surfaces, HUXLEY (1958); KALCKAR (1965). The aberrant sociology of

[1] Genetic mutations and the epigenetic introduction of a new membrane protein, coded by a tumor virus, should have similar consequences.

tumor cells, their invasive and often accelerated growth are thus distal to the crucial proximal event — the appearance of an aberrant protein into the surface and perhaps other membranes. Many of the distal effects may have trivial consequences, but others such as a possible increased affinity constant in a membrane transport system, SHRIVASTAVA and QUASTEL (1962), could give the altered cells a competitive advantage over normal and other mutant cells leading to the well-known phenomenon of tumor "progression".

In conclusion, the present hypothesis postulates that an oncogenic agent acts to introduce an inappropriate protein into cell mebranes (either in replacement of or in addition to normal components). Numerous membrane functions may be modified as a consequence, accounting for the membrane changes and the pleiomorphism seen in neoplasia. For example, the presence of an abnormal, structural component could result directly in morphological changes, altered electrokinetic properties, altered contact relationships, etc. On the other hand, there may also be more distal effects, such as altered binding of a variety of substances, including mRNA, various intermediates of metabolism and various lipids, and modifications of the function of otherwise normal, membrane-associated enzymes, in a manner analogous to that observed in *Neurospora*. Finally, one would expect to find defects, which, being several steps removed from the proximal event have no immediately obvious relationship to membranes. The abnormal control of glycolysis and of fatty acid synthesis might fall into this category. One would anticipate that the extent of the derangement caused by replacement of a normal membrane protein would depend upon its function and its position in the membrane lattice, and alterations of membrane "structural" proteins would have particularly far-reaching effects, especially if they are present in all membranes.

Finally, one should give consideration to the possible role of membranes in the initiation of the neoplastic process. This matter has been considered in some detail by von ARDENNE and RIEGER (1968).

Acknowledgements

Supported by Grant # CA 07382 of the U.S. Public Health Service. The author is the Andres Soriano Investigator of the Massachusetts General Hospital, Boston, Massachusetts.

References

ABELEV, G. I.: Antigenic structure of chemically-induced hepatomas. Progr. exp. Tumor Res. (Basel) **7**, 104—157 (1965).

ABERCROMBIE, M., and E. J. AMBROSE: Interference microscope studies of cell contacts in tissue culture. Exp. Cell Res. **15**, 332—345 (1958).

— — The surface properties of cancer cells: a review. Cancer Res. **22**, 525—548 (1962).

—, and J. E. M. HEAYSMAN: Observations on the social behavior of cells in tissue culture. II. 'monolayering' of fibroblasts. Exp. Cell Res. **6**, 293—306 (1954).

— —, and H. M. KARTHAUSER: Social behavior of cells in tissue culture. III. mutual influence of sarcoma cells and fibroblasts. Exp. Cell Res. **13**, 276—291 (1957).

ALEXANDER, P.: The role of tumor-specific antigens in the genesis; development and control of malignant disease. In: The biology of cancer (E. J. AMBROSE and F. J. C. ROW, eds.). New York: Van Nostrand 1966.

AMBROSE, E. J., A. M. JAMES, and J. H. B. LOWICK: Differences between the electrical charge carried by normal and homologous tumor cells. Nature (Lond.) **177**, 576—577 (1956).

ARDENNE, M. v., and F. RIEGER: Zu den Wegen der Krebstherapie und möglichen Ursachen der Krebsentstehung sowie ihren wechselseitigen Beziehungen. Z. Naturforsch. **23**B, 357—368 (1968).

ARONSON, A.: Membrane-bound messenger RNA and polysomes in sporulating bacteria. J. molec. Biol. **13**, 92—104 (1965).

ATKINSON, D. E.: Biological feedback control at the molecular level. Science **150**, 851—857 (1965).

ATTARDI, B., and G. ATTARDI: A membrane-associated RNA of cytoplasmic origin in Hela cells. Proc. nat. Acad. Sci. (Wash.) **58**, 1051—1058 (1967).

AUB, J., C. TIEFLAU, and A. LANKESTER: Reactions of normal and tumor cell surfaces to enzymes. I. Wheat-germ lipase and associated mucopolysaccharides. Proc. nat. Acad. Sci. (Wash.) **50**, 613—619 (1963).

AZOULAY, E., J. PUIG, and F. PICHINOTY: Alteration of respiratory particles by mutations in *Escherichia coli* K 12. Biochem. biophys. Res. Commun. **27**, 270—274 (1967).

BARSKI, G. E., and J. BELEHRADEK, JR.: Etude microcinematographique du mecanisme d'invasion cancereuse en cultures de tissue normal associe aux cellules malignes. Exp. Cell Res. **37**, 464—480 (1965).

BOREK, C., and L. SACHS: The difference in contact inhibition of cell replication between normal cells and cells transformed by different carcinogens. Proc. nat. Acad. Sci. (Wash.) **56**, 1705—1711 (1966).

BOXER, G. E., and T. M. DEVLIN: Pathways of intracellular hydrogen transport. Science **134**, 1495—1501 (1961).

BRIERLY, G., and D. E. GREEN: Compartmentation of the mitochondrion. Proc. natl. Acad. Sci. (Wash.) **53**, 73—79 (1965).

BURGER, M. M., and A. R. GOLDBERG: Identification of a tumorspecific determinant on neoplastic cell surfaces. Proc. natl. Acad. Sci. (Wash.) **57**, 359—366 (1967).

CASPAR, D. L. D.: Proceedings of the symposium on biophysics and physiology of biological transport (Frascati, 1965) eds., L. BOLIS, V. CAPRARO, K. R. PORTER, and J. D. ROBERTSON. Berlin-Heidelberg-New York: Springer 1967.

CHANCE, B., and B. HESS: Spectroscopic evidence of metabolic control. Science **129**, 700—708 (1959).

CHANGEUX, J. P., and T. R. PODLESKI: On the excitability and cooperativity of the electropax membrane. Proc. nat. Acad. Sci. (Wash.) **59**, 944—950 (1968).

—, and J. THIERY: In: Regulatory functions of biological membranes (BBA Library, No. 11), ed. J. JÄRNEFELT. Amsterdam: Elsevier 1968.

— — Y. TUNG, and C. KITTEL: On the cooperativity of biological membranes. Proc. nat. Acad. Sci. (Wash.) **57**, 334—341 (1967).

CHAPMAN, D., V. B. KAMAT, J. DEGIER, and S. A. PENKETT: Nuclear magnetic resonance studies of erythrocyte membranes. J. molec. Biol. **31**, 101—114 (1968).

— —, and R. J. LEVENE: Infrared spectra and the chain organization of erythrocyte membranes. Science **160**, 314—316 (1968).

COMAN, D. R.: Decreased mutual adhesiveness, a property of cells from squamous cell carcinomas. Cancer Res. **4**, 625—629 (1944).

— Mechanisms responsible for the origin and distribution of bloodborne tumor metastases; a review. Cancer Res. **13**, 397—404 (1953).

COMAN, D. R.: Cellular adhesiveness in relation to the invasiveness of cancer: electron microscopy of liver perfused with a chelating agent. Cancer Res. **14**, 519—521 (1954).

— Reduction in cellular adhesiveness upon contact with a carcinogen. Cancer Res. **20**, 1202—1204 (1960).

— Adhesiveness and stickiness; two independent properties of the cell surface. Cancer Res. **21**, 1436—1438 (1961).

—, and T. F. A. ANDERSON: A structural difference between the surface of normal and of carcinomatous epidermal cells. Cancer Res. **15**, 541—543 (1955).

COMINGS, G. E., and G. KAKEFUDA: Initiation of deoxyribonucleic acid replication at the nuclear membrane in human cells. J. molec. Biol. **33**, 225—229 (1968).

COOK, G. M. W., and W. JACOBSON: The electrophoretic mobility of normal and leukaemic cells of mice. Biochem. J. **107**, 549—557 (1968).

CURTIS, A. S. G.: Cell contacts: some physical considerations Amer. Naturalist **94**, 37—56 (1960).

— The cell surface. New York, London: Academic Press 1967.

DALTON, A. J.: An electron-microscopical study of a series of chemically induced hepatomas. In: Cellular control mechanisms and cancer, eds. P. EMMELOT and O. MUEHLBOCK. New York: Elsevier 1964.

DANIELLI, J. F., and H. DAVSON: A contribution to the theory of permeability of thin films. J. cell. comp. Physiol. **5**, 495—508 (1935).

ELLWORTH, J. C., Y. C. LIN, V. J. CRISTOFALO, S. WEINHOUSE, and H. P. MORRIS: Glucose utilization in homogenates of the Morris Hepatoma 5123 and related tumors. Cancer Res. **23**, 906—913 (1963).

FORESTER, J. A., E. J. AMBROSE, and J. A. MACPHERSON: Electrophoretic investigations of a clone of hamster fibroblasts and polyoma-transformed cells from the same population. Nature (Lond.) **196**, 1068—1070 (1952).

— —, and M. G. P. STOKER: Microelectrophoresis of normal and transformed clones of hamster kidney fibroblasts. Nature (Lond.) **201**, 945—946 (1964).

—, and M. H. SALAMANN: Cell electrophoretic mobilities in Friend virus disease. Nature (Lond.) **215**, 279—280 (1967).

FRITZ, I. B.: An hypothesis concerning the role of carnitine in the control of interrelations between fatty acid and carbohydrate metabolism. Perspect. Biol. Med. **10**, 643—677 (1967).

—, and M. P. HSU: Studies on the control of fatty acid synthesis. J. biol. Chem. **242**, 865—872 (1967).

FURSHPAN, E. I., and D. D. POTTER: Low-resistance junctions between cells in embryos and tissue culture. In: Current topics in developmental biology. New York: Academic Press 1968.

GAITSKHOKI, V. S.: Distribution of adenylic coenzymes between mitochondria and hyaloplasm as a glycolysis controlling factor in the hepatic cell of mouse and of Ehrlich ascites carcinoma. Biokhimiya **26**, 926—933 (1961).

— Role of intramitochondrial coenzymes in the control of the rate of glycolysis. Biokhimiya **27**, 286—292 (1962).

GANESAN, A. T., and J. LEDERBERG: A cell-membrane bound fraction of bacterial DNA. Biochem. biophys. Res. Commun. **18**, 824—835 (1965).

GANSCHOW, R., and K. PAIGEN: Separate genes determining the structure and intracellular location of hepatic glucuronidase. Proc. natl. Acad. Sci. (Wash.) **5**, 938—945 (1967).

GOLD, P., and S. O. FREEDMAN: Demonstration of tumor-specific antigens in human colonic carcinomata by immunological tolerance and absorption techniques. J. exp. Med. **121**, 439—462 (1965a).

— — Specific carcinoembryonic antigens of the human digestive system. J. exp. Med. **122**, 467—481 (1965b).

GOLD, P., M. GOLD, and S. O. FREEDMAN: Cellular location of carcinoembryonic antigens of the human digestive system. Cancer Res. **28**, 1331—1334 (1968).

GRAFFI, A., G. BUTSCHAK, and E. J. SCHNEIDER: Differences of mitochondrial protein synthesis *in vitro* between tumor and normal tissues. Biochem. biophys. Res. Commun. **21**, 418—423 (1965).

GREEN, D. E., and S. FLEISCHER: The role of lipids in mitochondrial electron transfer and oxidative phosphorylation. Biochim. biophys. Acta (Amst.) **70**, 554—582 (1963).

HAKOMORI, S., J. KOSCIELAK, K. J. BLOCH, and R. W. JEANLOZ: Immunologic relationship between blood group substances and a fucose-containing glycolipid of human adenocarcinoma. J. Immunol. **98**, 31 (1967).

—, and W. MURAKAMI: Glycolipids of hamster fibroblasts and derived malignant-transformed cell lines. Proc. nat. Acad. Sci. (Wash.) **59**, 254—261 (1968).

HALPERN, B., B. PEJSACHOWICZ, H. L. FEBVRE, and G. BARSKI: Differences in patterns of aggregation of malignant and nonmalignant mammalian cells. Nature (Lond.) **209**, 157—159 (1966).

HARDESTY, B. A., and H. K. MITCHELL: The accumulation of free fatty acids in Poky, a maternally inherited mutant of Neurospora crassa. Arch. Biochem. **100**, 330—334 (1963).

HARRIS, R. J. C.: Specific tumor antigens. New York: Medical Examination Publ. Co. 1967.

HART, L., R. H. ADAMSON, H. P. MORRIS, and J. R. FOUTS: The stimulation of drug metabolism in various rat hepatomas. J. Pharmacol. exp. Ther. **149**, 7—15 (1965).

HASKINS, F. A., A. TISSIERES, H. K. MITCHELL, and M. B. MITCHELL: Cytochromes and the succinic acid oxidase system of Poky strains of Neurospora. J. biol. Chem. **200**, 819—826 (1953).

HELLSTROM, K. E., and G. MOLLER: Immunological and immunogenetic aspects of tumor transplantation. Progr. Allergy **9**, 158—245 (1965).

HILL, B. R.: Some properties of serum lactic dehydrogenase. Cancer Res. **16**, 460—467 (1956).

HOLMBERG, B.: On the *in vitro* release of cytoplasmic enzymes from ascites tumor cells as compared with strain L cells. Cancer Res. **21**, 1386—1393 (1961).

HULTIN, H. O., and S. H. RICHARDSON: The binding of phosphate, pyrophosphate, and nucleotides to the structural protein of beef heart mitochondria. Arch. Biochem. **105**, 288—296 (1964).

HUXLEY, J.: Biological aspects of cancer. New York: Harcourt, Brace & Co. 1958.

JACOB, F., F. BRENNER and S. CUZIN: On the regulation of DNA replication in bacteria. In: Cold Spring Harbor symposium on synthesis and structure of macromolecules, vol. 28. New York: Cold Spring Harbor Lab. of Quant. Biol. 1963.

— A. RYTER, and F. CUZIN: On the association between DNA and membrane in bacteria. Proc. roy. Soc. B **164**, 267—278 (1966).

JAMAKOSMANOVIC, A., and W. R. LOEWENSTEIN: Cellular uncoupling in cancerous thyroid epithelium. Nature (Lond.) **218**, 775 (1968).

JOHNSON, M. J.: The role of aerobic phosphorylation in the Pasteur effect. Science **94**, 200—202 (1941).

KALCKAR, H. M.: Galactose metabolism and cell "society". Science **150**, 305—313 (1965).

KANNO, Y., and Y. MATSUI: Cellular uncoupling in cancerous stomach epithelium. Nature (Lond.) **218**, 775—776 (1968).

KLEIN, G.: Tumor antigens. In: The specificity of cell surfaces, eds. B. D. DAVIS and L. WARREN. New Jersey: Prentice-Hall, Inc. 1967.

— Tumor-specific transplantation antigens: G. H. A. Clover Memorial Lecture. Cancer Res. **28**, 625—635 (1968).

KLINGENBERG, M., and E. PFAFF: Structural and functional compartmentation in mitochondria. In: Regulation of metabolic processes in mitochondria (BBA Library No 7), eds. J. M. TAGER, F. PAPA, E. QUAGLIORIELLO, and E. C. SLATER. New York: Elsevier 1966.

KORN, E. D.: Structure of biological membranes. Science **153**, 1491—1498 (1966).

LAMAR, C. JR., M. PRIVAL, and H. C. PITOT: Studies on the mechanism of template stability. I. The effect of actinomycin on the base composition and sedimentation of ^{32}P-labeled RNA in rat liver. Cancer Res. **26**, 1909—1914 (1966).

LARK, C., and K. G. LARK: Evidence for two distinct aspects of the mechanism regulating chromosome replication in escherichia coli. J. molec. Biol. **10**, 120—136 (1964).

LARK, K. G.: Regulation of chromosome replication and segregation in bacteria. Bact. Rev. **30**, 3—32 (1966).

LEHNINGER, A. L.: The neuronal membrane. Proc. natl. Acad. Sci. (Wash.) **60**, 1069—1080 (1968).

LENARD, J., and S. J. SINGER: Protein conformation in cell membrane preparations as studied by optical rotatory dispersion and circular dichroism. Proc. natl. Acad. Sci. (Wash.) **56**, 1828—1835 (1966).

— — Structure of membranes: reaction of red blood cell membranes with phospholipase C. Science **159**, 738—739 (1968).

LEVINE, D., Y. BECKER, C. BOONE, and H. EAGLE: Contact inhibition macromolecular synthesis, and polyribosomes in cultured human diploid fibroblasts. Proc. nat. Acad. Sci. (Wash.) **53**, 350—356 (1965).

LIN, Y. P., J. C. ELLWORTH, A. ROSADO, H. P. MORRIS, and S. WEINHOUSE: Glucose metabolism in a low-glycolysing tumor, the Morris Hepatoma 5123. Nature (Lond.) **195**, 153—155 (1962).

LO, C.-H., V. J. CRISTOFALO, H. P. MORRIS, and S. WEINHOUSE: Studies on respiration and glycolysis in transplanted hepatic tumors of the rat. Cancer Res. **28**, 1—10 (1968).

LOEWENSTEIN, W. R.: Permeability of membrane junctions. Ann. N.Y. Acad. Sci. **137**, 441—472 (1966).

—, and Y. KANNO: Intercellular communication and tissue growth. I. Cancerous growth. J. Cell Biol. **33**, 225—234 (1967).

—, and R. D. PENN: Intercellular communication and tissue growth. II. Tissue regeneration. J. Cell Biol. **33**, 235—242 (1967).

— S. J. SOCOLAR, Y. KANNO, and N. DAVIDSON: Intercellular communication: renal, urinary bladder, sensory, and salivary gland cells. Science **149**, 295—298 (1965).

LOWICK, J. H. B., L. PURDOM, A. M. JAMES, and E. J. AMBROSE: Some microelectrophoretic studies of normal and tumor cells. J. roy. micr. Soc. **80**, 47—57 (1961).

LYNEN, F.: The aerobic phosphate requirements of yeast. The Pasteur reaction. Ann. Chem. **546**, 120—141 (1941).

MACPHERSON, I., and M. STOKER: Polyoma transformation of hamster cell clones — an investigation of genetic factors affecting cell competence. Virology **16**, 147—151 (1962).

MADDY, A. H.: The chemical organization of the plasma membrane of animal cells. Int. Rev. Cytol. **20**, 1—65 (1966).

—, and B. R. MALCOLM: Protein conformations in the plasma membrane. Science **150**, 1616—1618 (1965).

— — Protein conformations in biological membranes. Science **153**, 213 (1966).

MAJERUS, J. S. M., R. JACOBS, and M. B. SMITH: The regulation of fatty acid biosynthesis in rat hepatoma. J. biol. Chem. **243**, 3588—3595 (1968).

Makita, A., C. Suzuki, and A. Yosizawa: Chemical and immunological characterization of the Forssman hapten isolated from equine organs. J. Biochem. (Tokyo) **60**, 502—513 (1966).

—, and T. Yamakawa: Biochemistry of organ glycolipids. I. Ceramide-oligonexosides of human, equine and bovine spleens. J. Biochem. (Tokyo) **51**, 124—133 (1962).

Malmgren, H., B. Sylven, and L. Revesz: Catheptic and dipeptidase activities of ascites tumor cells. British J. Cancer **9**, 473—479 (1955).

Masamune, H., Z. Yosizawa, P. Oh-Uti, J. Matsuda, and A. Masuka: Biochemical studies on carbohydrates. Tohoku J. exp. Med. **56**, 37—42 (1952).

McNair-Scott, T. B., K. K. Sanford, and B. B. Westfall: Growth enzyme activities correlated with tumor-producing capacity of four cell lines derived from one cell. Proc. Amer. Ass. Cancer Res. **3**, 41 (1959).

Mitchell, M. B., and H. K. Mitchell: A case of "maternal" inheritance in Neurospora crassa. Proc. natl. Acad. Sci. (Wash.) **38**, 442—449 (1952).

Moor, H.: Use of freeze-etching in the study of biological ultrastructure. In: International review of experimental pathology, eds. G. W. Richter and M. A. Epstein. New York: Academic Press 1966.

Morris, H. P.: Studies on the development, biochemistry, and biology of experimental hepatomas. Advanc. Cancer Res. **9**, 227—302 (1965).

Munkres, K. D., and D. O. Woodward: On the genetics of enzyme locational specificity. Proc. nat. Acad. Sci. (Wash.) **55**, 1217—1224 (1966b).

Nachmansohn, D.: Chemical control of the permeability cycle in excitable membranes during electrical activity. Ann. N.Y. Acad. Sci. **137** (2), 877—900 (1966).

Nakas, M., S. Higashino, and W. R. Loewenstein: Uncoupling of an epithelial cell membrane junction by calcium-ion removal. Science **151**, 89—91 (1966).

Neifakh, S. A., T. B. Kazakova, M. P. Melnikova, and V. F. Turovskii: The membrane mechanism of regulation of glycolysis rate in cells. Dokl. Akad. Nauk. SSSR **138**, 227 (1961).

Nilausen, K., and H. Green: Reversible arrest of growth in G1 of an established fibroblast line (3T3). Exp. Cell Res. **40**, 166—168 (1965).

Numa, S., W. M. Bortz, and F. Lynen: Regulation of fatty acid synthesis at the acetyl-CoA carbioxylation step. Advanc. Enzyme Reg. **3**, 407—423 (1965).

Ohe, K., H. P. Morris, and S. Weinhouse: β-hydroxybutyrate dehydrogenase activity in liver and liver tumors. Cancer Res. **27**, 1360—1371 (1967).

Ohta, N., A. B. Pardee, B. R. McAuslan, and M. M. Burger: Sialic acid contents and controls of normal and malignant cells. Biochim. biophys. Acta (Amst.) **158**, 98—102 (1968).

Okada, Y.: Analysis of giant polynuclear cell formation caused by HVJ virus from Ehrlich ascites tumor cells. I. Microscopic observation of giant polynuclear cell formation. Exp. Cell Res. **26**, 98—107 (1962).

—, and F. Murayama: Requirement of calcium ions for the cell fusion reaction of animal cells by HVJ. Exp. Cell Res. **44**, 527—551 (1966).

— —, and K. Yamada: Requirement of energy for the cell fusion reaction of Ehrlich ascites tumor cells by HVJ. Virology **27**, 115—130 (1966).

—, and J. Tadokoro: Analysis of giant polynuclear cell formation caused by HVJ virus from Ehrlich ascites tumor cells. II. Quantitative analysis of giant polynuclear cell formation. Exp. Cell Res. **26**, 108—118 (1962).

— — The distribution of cell fusion capacity among several cell strains or cells caused by HVJ. Exp. Cell Res. **32**, 417—430 (1963).

Old, L. J., and E. A. Boyse: Immunology of experimental tumor. Ann. Rev. Med. **15**, 167—186 (1964).

O'Neill, C. H.: An association between viral transformation and Forssman antigen detected by immune adherence in cultured BHK21 cells. J. Cell Sci. **3** (3), 405—421 (1968).

PENN, R. D.: Ionic communication between liver cells. J. Cell Biol. **29**, 171—174 (1966).

PERUTZ, M. F.: Structure and function of haemoglobin. I. A tentative atomic model of horse oxyhaemoglobin. J. molec. Biol. **13**, 646—668 (1965).

— J. D. KENDREW, and H. C. WATSON: Structure and function of haemoglobin. II. Some relations between polypeptide chain configuration and amino acid sequence. J. molec. Biol. **13**, 669—678 (1965).

PITOT, H. C.: Altered template stability: the molecular mask of malignancy? Perspect. Biol. Med. **8**, 50—70 (1964).

— Some biochemical aspects of malignancy. Ann. Rev. Biochem. **35** (I), 335—368 (1966).

—, and Y. S. CHO: Control mechanisms in the normal and neoplastic cell. Progr. exp. Tumor Res. **7**, 158—223 (1965).

— C. PERAINO, C. LAMAR, JR., and A. L. KENNAN: Template stability of some enzymes in rat liver and hepatoma. Proc. nat. Acad. (Wash.) **54**, 845—851 (1965).

POTTER, D. D.: Personal communication.

— E. I. FURSHPAN, and E. J. LENNOX: Connections between cells of the developing squid as revealed by electrophysiological methods. Proc. nat. Acad. Sci. (Wash.) **55**, 328—335 (1966).

PURDOM, L., E. J. AMBROSE, and G. E. KLEIN: A correlation between electrical surface charge and some biological characteristics during the stepwise progression of a mouse sarcoma. Nature (Lond.) **181**, 1586—1587 (1958).

RACKER, E.: Regulation of adenosine triphosphate utilization in multi-enzyme systems. In: Mechanisms in bioenergetics. New York: Academic Press 1965.

RAUCH, H.: Electrophoretic membrane protein variants associated with D-locus in the house mouse. Genetics **60**, 214 (1968).

ROBERTSON, J. D.: The ultrastructure of cell membranes and their derivatives. In: Biochemical society symposia; 16. The structure and function of subcellular components, ed. E. M. CROOK. Cambridge, England: The University Press 1959.

ROSENBERG, M. D.: Single cell properties-membrane development. In: Ciba foundation symposium on cell differentiation, eds. A. V. S. DEREUCK, and J. KNIGHT. London, England: J. & A. Churchill, Ltd. 1967.

RYSER, H.: Uptake of protein by mammalian cells: an underdeveloped area. Science **159**, 390—396 (1968).

RYTER, A.: Association of the nucleus and the membrane of bacteria: a morphological study. Bacter. Rev. **32**, 39—54 (1968).

—, and F. JACOB: Etude au microscope electronique des relations entre mesosomes et noyaux chez bacillus subtilis. C. R. Acad. Sci. (Paris) **257**, 3060—3063 (1963).

RUBIN, H.: The inhibition of chick embryo cell growth by medium obtained from cultures of rous sarcoma cells. Exp. Cell Res. **41**, 149—161 (1966).

SABINE, J. R., S. ABRAHAM, and I. L. CHAIKOFF: Lack of feedback control of fatty acid synthesis in a transplantable hepatoma. Biochim. biophys. Acta (Amst.) **116**, 407—409 (1966).

— — — Control of lipid metabolism in hepatoma: insensitivity of rate of fatty acid and cholesterol synthesis by mouse hepatoma BW 7756 to fasting and to feedback control. Cancer Res. **27**, 793—799 (1967).

— —, and H. P. MORRIS: Defective dietary control of fatty acid metabolism in four transplantable rat hepatomas: numbers 5123C, 7793, 7795, and 7800. Cancer Res. **28**, 46—51 (1968).

SACHS, L., and D. MEDINA: In vitro transformation of normal cells by polyoma virus. Nature (Lond.) **189**, 457—458 (1961).

SHAPIRO, B.: Lipid metabolism. Ann. Rev. Biochem. **36** (I), 247—270 (1967).

SHRIVASTAVA, G. C., and J. H. QUASTEL: Malignancy and tissue metabolism. Nature (Lond.) **196**, 876—880 (1962).

SIMON-REUSS, I., G. M. W. COOK, G. V. F. SEAMAN, and D. H. HEARD: Electrophoretic studies on some types of mammalian tissue cell. Cancer Res. **24**, 2038—2043 (1964).

SIPERSTEIN, M. D., and V. M. FAGAN: Deletion of the cholesterol-negative feedback system in liver tumors. Cancer Res. **24**, 1108—1115 (1964).

— — Feedback control of mevalonate synthesis by dietary cholesterol. J. biol. Chem. **24**, 602—609 (1966).

— —, and H. P. MORRIS: Further studies on the deletion of the cholesterol feedback system in hepatomas. Cancer Res. **26**, 7—11 (1966).

—, and M. J. GUEST: Studies on the site of the feedback control of cholesterol synthesis. J. clin. Invest. **39**, 642—652 (1960).

SJOGREN, H. O.: Transplantation methods as a tool for detection of tumor-specific antigens. Progr. exp. Tumor Res. **6**, 289—322 (1965).

SJOSTRAND, F.: Molecular structure and function of cellular mambranes. In: 15th colloquium on protides of the biological fluids, ed. H. PEETERS. New York: Elsevier 1968.

SPECTOR, A. A.: The importance of free fatty acid in tumor nutrition. Cancer Res. **27**, 1580—1586 (1967).

—, and D. STEINBERG: The effect of fatty acid structure on utilization by Ehrlich ascites tumor cells. Cancer Res. **27**, 1587—1594 (1967).

STEIM, J.: Studies of biological membrane structure by thermal and spectroscopic methods, 153rd Meeting of Amer. Chem. Soc. (Miami Beach, April) 1967.

STOKER, M.: Regulation of growth and orientation in hamster cells transformed by polyoma virus. Virology **24**, 165—174 (1964).

— Contact and short-range interactions affecting growth of animal cells in culture. In: Current topics in developmental biology, vol. 2. New York: Academic Press 1967a.

— Transfer of growth inhibition between normal and virus-transformed cells; autoradiographic studies using marked cells. J. Cell Sci. **2**, 293—304 (1967b).

— M. SHEARER, and C. O'NEILL: Growth inhibition of polyoma-transformed cells by contact with static normal fibroblasts. J. Cell Sci. **1**, 297—310.

SUBAK-SHARPE, H.: Biochemically marked variants of the Syrian hamster fibroblast cell line BHK21 and its derivatives. Exp. Cell Res. **38**, 106—119 (1965).

SUGIMURA, T., K. IKEDA, K. HIROTA, M. HOZEMI, and H. P. MORRIS: Chemical, enzymatic, and cytochrome assays of microsomal fraction of hepatomas with different growth rates. Cancer Res. **26**, 1711—1716 (1966).

SUSS, R., G. BLOBEL, and H. C. PITOT: Rat liver and hepatoma polysome-membrane interaction *in vitro*. Biochem. biophys. Res. Commun. **23**, 299—304 (1966).

SYLVEN, B.: The biochemical mechanism underlying the destructive growth of tumor. Acta Un. int. Cancr. **14**, 61—62 (1958).

— The host-tumor interzone and tumor invasion. In: Biological interactions in normal and neoplastic growth: Henry Ford Hospital Internat. Symposium, eds. M. J. BRENNAN and W. L. SIMPSON. Boston, Mass.: Little, Brown & Co. 1962.

—, and I. BOIS: Protein content and enzymatic assays of interstitial fluid from some normal tissues and transplanted mouse tumors. Cancer Res. **20**, 831—836 (1960).

—, and H. MALMGREN: Topical distribution of proteolytic activities in some transplanted mice tumor. Exp. Cell Res. **8**, 575—577 (1955).

— — The histological distribution of preoteinase and peptidase activity in solid tumor transplants. Acta radiolog. (Stockh.) Suppl. **154**, 1—124 (1957).

— R. OTTOSON, and L. REVESZ: The content of dipeptidases and acid proteinases in the ascitic fluid of mice with ascites tumor. Brit. J. Cancer **13**, 551—565 (1959).

TEMIN, H. M., and H. RUBIN: Characteristics of an assay for rous sarcoma virus and rous sarcoma cells in tissue culture. Virology **6**, 669—688 (1958).

TISSIERES, A., and H. K. MITCHELL: Cytochromes and respiratory activities in some slow growing strains of Neurospora. J. biol. Chem. **208**, 241—249 (1954).
— —, and F. A. HASKINS: Studies on the respiratory system of the Poky strain of Neurospora. J. biol. Chem. **205**, 423—433 (1953).
TODARO, G., H. GREEN, and B. GOLDBERG: Transformation of properties of an established cell line by SV_{40} and polyoma virus. Proc. nat. Acad. Sci. (Wash.) **51**, 66—73 (1964).
— K. HABEL, and H. GREEN: Antigenic and cultural properties of cell doubly transformed by polyoma virus and SV_{40}. Virology **27**, 179—185 (1965).
— G. LAZAR, and H. GREEN: The initiation of cell division in a contact-inhibited mammalian cell line. J. cell. comp. Physiol. **66**, 325—333 (1966).
TOMKINS, G. M., H. SHEPPARD, and I. L. CHAIKOFF: Cholesterol synthesis by liver. III. Its regulation by ingested cholesterol. J. biol. Chem. **201**, 137—141 (1953).
TUPPY, H., P. SWETLEY, and I. WOLFF: Structural proteins of wild-type (p^+) and respiration-deficient (p^-) yeast mitochondria: a comparative study using electrophoretic and immunological methods. Europ. J. Biochem. **968**, 339—345 (1968).
URRY, D. W., M. MEDNIEKS, and E. BEJNAROWICZ: Optical rotation of mitochondrial membranes. Proc. nat. Acad. Sci. (Wash.) **57**, 1043—1049 (1967).
VASSAR, P. S.: The electric charge density of human tumor cell surfaces. Lab. Invest. **12**, 1072—1077 (1963).
VOGEL, M., and L. SACHS: Studies on the antigenic composition of hamster tumors induced by polyoma virus, and of normal hamster tissues *in vivo* and *in vitro*. J. nat. Cancer Inst. **29**, 239—252 (1962).
— — The induction of Forssman-antigen synthesis in hamster and mouse cells in tissue culture, as detected by the fluorescent-antibody technique. Exp. Cell Res. **34**, 448—462 (1964).
VOGT, M., and R. DULBECCO: Virus-cell interaction with a tumor-producing virus. Proc. nat. Acad. Sci. (Wash.) **46**, 365—370 (1960).
WALLACH, D. F. H.: Isolation of plasma membranes of animal cells. In: Specificity of cell surfaces, eds. B. D. DAVIS and L. WARREN. Englewood Cliffs, New Jersey: Prentice-Hall, Inc. 1967.
—, and A. GORDON: Lipid-protein interaction in cellular membranes. In: Regulatory functions of biological membranes (BBA Library, No. 11), ed. J. JÄRNEFELT. Amsterdam: Elsevier 1968.
— — Protein conformations in cellular membranes. In: 15th Colloquium on Protides of the Biological Fluids, ed. H. PEETERS. New York: Elsevier 1968a.
— — Lipid protein interactions in cellular membranes. Fed. Proc. **21** (6), 1263—1268 (1968b).
—, and V. B. KAMAT: The contribution of sialic acid to the surface charge of fragments of plasma membrane and endoplasmic reticulum. J. Cell Biol. **30**, 660—663 (1966).
— —, and M. H. GAIL: Physiochemical differences between fragments of plasma membrane and endoplasmic reticulum. J. Cell. Biol. **30**, 601—621 (1966).
—, and M. V. PEREZ-ESANDI: Sialic acid and the electrophoretic mobility of three tumor cell types. Biochim. biophys. Acta (Amst.) **83**, 363—366 (1964).
—, and V. VLAHOVIC: Immunological cross-reactivity between plasma membrane and endoplasmic reticulum of Ehrlich ascites carcinoma. Nature (Lond.) **216**, 182 (1967).
—, and H. P. ZAHLER: Protein conformations in cellular membranes. Proc. nat. Acad. Sci. (Wash.) **56**, 1552—1559 (1966).
— — Infrared spectra of plasma membrane and endoplasmic reticulum of Ehrlich ascites carcinoma. Biochim. biophys. Acta (Amst.) **150**, 186—193 (1968).
WATANABE, A., I. TASAKI, and L. LERMAN: Bi-ionic action potentials in squid giant axons internally perfused with sodium salts. Proc. nat. Acad. Sci. (Wash.) **58**, 2246—2252 (1967).

WEBB, T. E., G. BLOBEL, and V. R. POTTER: Polyribosomes in rat tissues. I. A study of *in vitro* patterns in liver and hepatomas. Cancer Res. **24**, 1229—1237 (1964).

— — —, and H. P. MORRIS: Polyribosomes in rat tissues. II. The polyribosome distribution in the minimal deviation hepatomas. Cancer Res. **25**, 1219—1224 (1965).

WEBER, G., G. BANERJEE, and H. P. MORRIS: Comparative biochemistry of hepatomas. I. Carbohydrate enzymes in Morris hepatoma 5123. Cancer Res. **21**, 933—937 (1961).

WEINSTEIN, R. S., and V. M. KOO: Penetration of red cell membranes by some membrane-associated particles. Proc. Soc. exp. Biol. (N.Y.) 1968 (in press).

—, and R. A. WILLIAMS: Freeze-cleaving of RBC membranes in paroxysmal nocturnal hemoglobinuria. Blood **30**, 186—191 (1967).

WEISS, L., and E. MAYHEW: The presence of ribonucleic acid within the peripheral zones of two types of mammalian cell. J. cell. comp. Physiol. **68**, 345—359 (1966).

WENNER, C. E.: Progress in tumor enzymology. Advanc. Enzymol. **29**, 321—390 (1967).

WIENER, J., D. SPIRO, and W. LOEWENSTEIN: Ultrastructure and permeability of nuclear membranes. J. Cell Biol. **27**, 107—117 (1965).

WOOD, W. A.: Carbohydrate metabolism. Ann. Rev. Biochem. **35**, 521—558 (1966).

WOODWARD, D. O.: Structural protein and membrane formation. Fed. Proc. **27** (5), 1167—1173 (1968).

—, and K. D. MUNKRES: Alterations of a maternally inherited mitochondrial structural protein in respiratory-deficient strains of Neurospora. Proc. nat. Acad. Sci. (Wash.) **55**, 872—880 (1966a).

— — See MUNKRES and WOODWARD (1966b).

— — Genetic control, function, and assembly of a 'structural' protein in *Neurospora*. In: Organizational biosynthesis, eds. H. J. VOGL, J. O. LAMPEN, any V. BRYSON. New York: Academic Press 1967.

WROBLEWSKI, F.: The mechanisms of alteration in lactic dehydrogenase activited of body fluids. Ann. N. Y. Acad. Sci. **75**, 322—338 (1958).

WU, H., E. MEEZAN, P. H. BLACK, and P. W. ROBBINS: Differences in the composition of particulate fractions of mouse fibroblast 3T3 and SV_{40} transformed 3T3. Fed. Proc. **27**, (2) 814 (1968).

WU, R.: Leakage of enzyme from ascites tumor cells. Cancer Res. **75**, 1217—1222 (1959).

—, and E. RACKER: Leakage of enzymes from ascites tumor cells. Fed. Proc. **17**, 339 (1958).

WU, W.: "Minimal deviation" hepatomas; a critical review of the terminology including a commentary on the correlation of enzyme activity with growth rate of hepatomas. J. nat. Cancer Inst. **39**, 1149—1154 (1967).

Author Index

Page numbers in *italics* refer to the bibliography

Subject Index

www.ingramcontent.com/pod-product-compliance
Lightning Source LLC
Chambersburg PA
CBHW061208240926
51846CB00006B/1153
9783642461620